The Sheep of Dolgwili

The Sheep of Dolgwili

R. J. Fayers

HODDER AND STOUGHTON
LONDON SYDNEY AUCKLAND TORONTO

British Library Cataloguing in Publication Data
Fayers, R. J.
　The sheep of Dolgwili.
　1. Sheep – Wales
　I. Title
　636.3′0092′4　　sf375.7.g7
　ISBN 0 340 25532 3

Contents

The Sheep of Dolgwili

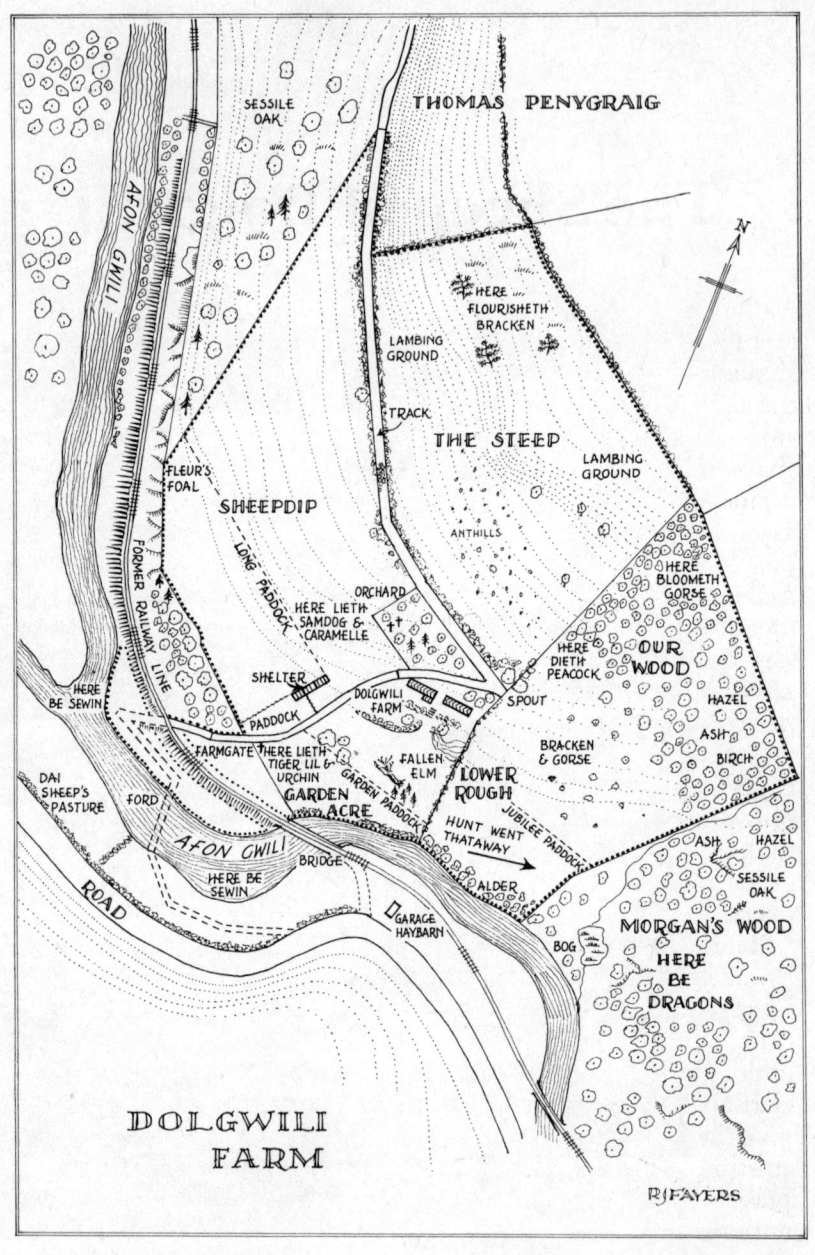

1   Brave words, Grandpa

Along the bottom of the dead elm branch, a barred woodpecker crept, upside down, pied and red, furiously pecking at and shaking off flakes of dead bark and moss. But along the top of the same branch a blurred green shape, out-of-focus, glided to a stop. A fingertip corrected the binocular focus; the blurred shape clarified to be a green bus. Off it dropped an old man and a small child – Dai Sheep, all gnarled oak, and his willow-sapling grand-daughter, intent on more sheep business. The bus laboured off down the valley road, the little girl went frisking off through the meadowsweet down into the low meadow beside the river, whilst Dai Sheep dragged his heavy feet down the sloping pathway after her, his bent balance already dependent on his hazel thumbstick, and made more precarious this evening by the heavy bottle swinging in his jacket pocket. The child, twelve years old perhaps, ran as eagerly as a sheepdog puppy towards the dozen sheep, sweetly oblivious to her grandfather's stiff-legged wrestle with the makeshift stile which threatened to make of his entrance to the green meadow stage below a toss-up between high farce and deep tragedy.

The motley sheep, used to solitude, stopped grazing, looked towards the running, waving child and began to run scared themselves. Dai Sheep, somehow arrived on ground as level as now he was ever likely to know, must have commanded the child, whilst he stood crookedly leaning his weight on his stick perhaps to consider the work on hand, to regain his breath and to give the rocking world time to settle, must have commanded her to stillness. He slowly lugged the heavy bottle from his pocket, peered at his alert and now stationary flock and then pointed with his thumbstick. That one . . . No, that one there, no doubt

9

he was instructing his infant shepherdess, for in she dived scattering the flock right, left and centre. As they fled, the child chased, up and down and round and round the little half-acre shelf of pasture, with old Dai brandishing his stick and certainly croaking angry instructions, an ancient general commanding a puny army of one in an unwinnable skirmish. Time after time the puppydog-girl and the brittle old man manoeuvred the handful of sheep towards a corner, at first approximately, then with increased definition, then stealthily, until the one wanted sheep was tantalisingly near to hand, when the flock would bolt and leap and pour out from their encompassment up along the meadow to resume their alert watching or their half-hearted grazing, until once again Little Bo-Peep came running towards them for one more tormenting merry-go-round.

The small green stage on which these puppets performed was no more than a shelf of poor pasture, long and narrow, set between main road and river at the bottom of a vast amphitheatre of tree-girt hills. Just above river-level and well below road-level, the pasture was of little use to anyone except Dai Sheep, who tended to collect such odd bits of land, doubtlessly for penny-rents, just as he tended to collect odd bits of sheep. They were, somehow, magnetised by each other – little pastures, old sheep, and Dai – and they led to such passes as this one. Dogless, penless, crookless, how could such an ancient and such an innocent expect to separate and catch just the one particular creature that the bottle-drench was intended for?

There was, however, the bare-boned ruin of a shelter in the meadow – a few timbers, a sheet or two of corrugated tin – but badly placed, up under the road hedge, on top of the steepest bank. Here, somehow, Bo-Peep had captured her sheep. Arms about his neck, she lay upon it and wrestled to hold it secure whilst fifteen feet below her, on level ground, grandfather called up to her, good girl, good girl now, to hold on now, he was coming up now, you see, good girl, coming up he was . . .

Brave words, Grandpa, but how? You all arthritic and within a sniff of death anyway, and that great bottle in your hand and all.

Dai Sheep set out on his Everest. His thumbstick unsteadily probed the summer grass, his old boots teased awkwardly at hummocks, as step by precarious step he began to climb. Up above, Bo-Peep held on. Dai inched upwards, gravity-stricken. He abandoned his thumbstick; he had need of an ice-pick and crampons. Bent fingers clawed hawthorn seedlings, grass, skyhooks, anything to pull that old body up to his sheep. Hold on, girl, he would be croaking, nearly there we are. And then, close to the top, it all fell apart and, old arms and stiff legs windmilling, the bent old man went sailing arse-over-tip all the rib-cracking way down the brutal slope, there painfully to roll like an old bough in a gale over on to his thin knees, to cast his faded eyes skyward and shake his sheep-wasted fists at a god who had so cursed him with this love of foolish sheep.

By now, of course, the sheep had struggled free. The petulant little girl sat scolding foolish Grandpa, who crawlingly retrieved his noteworthy hat to cover his white thatch, and then his great bottle of sheep-drench. Like some mechanical toy, he wound himself up, became nearly upright, and set off again.

For all of another evening hour they worked, until at last they caught their sheep at the lowest level available to them, down in the sweet-gliding and shallow river where, with Dai more or less astride it and holding its head, Bo-Peep upended the bottle and poured the drench down its throat. The evening show was over. All they had to do now was to haul themselves back on stage as it were. The sheep shook its head and stood blinking at man and child as they made their slow exit. Dai Sheep absently threw the empty bottle into the river and the child put her hand into her grandfather's. A pastoral idyll . . . Somehow they persuaded themselves back up on to the road, to begin the two-mile walk back to the village.

As I lowered the binoculars, my Temptress glided to my shoulder and cooed, "Well? If an octogenarian and a slip of a girl can manage sheep, what's stopping us, for goodness sake? We're only in our fifties . . . "

2　Right in the eye of the sun

The man from the ministry looked about him as he left his car, obviously wondering how on earth he got up to the place. He prowled around a bit before finding the hole in the hedge and thus the track that led him on to Dolgwili. He was the standard charmer, standing there smelling of central heating, listening with a tiny smile on his unweathered features while all the time looking about him at the unkempt hills. His smile seemed to suggest already that of course the idea of sheep-farming here was quite ludicrous for anyone so inexperienced as ourselves, but as he considered with faint interest the aspect of the hills about him, he asked only that he be allowed to wander around by himself and to report later.

Dolgwili, as Powell, our original Welsh mentor, had enthused before ever we bought the place, lay right in the eye of the sun. Twenty-three sweet acres of freedom, stretched across the hillside from the hedged horizon high above us down to the tinkling river, with a neat white cottage tucked in on a shelf halfway up the slope. Nice little wood, too, Powell had gently raved. Every farmer appreciated a nice little wood – stakes, logs for winter fires. It was good grass, too. All the pastures needed was close grazing and basic slag. Make your own hay, and the neighbours would all lend a hand with the haymaking and the shearing. And at foxglove time the river would be teeming with sewin and salmon. And so on, long into the night Powell had talked, selling us not merely Dolgwili, but the whole lovely Gwili valley and much of his beloved West Wales as he mapped out our entire sheep-farming future, without hearing a word of my sleepy protest that no, no, we were not going to farm, no fear we were not – ever.

No fear. Dolgwili was not for farming; it was for looking at, for walking over or for running a saluki. It was space, blessed wilderness, freedom. It was a retreat from suburbia, a hillside grandstand from which to consider with pity the poor capers of those still slogging it out in the inane race down there in the market place, a race from which we had already scratched. Anyway, it was far too steep, too small, too neglected and too untameable ever to farm. No. Dolgwili was for contemplation, for cloisters. A fine large garden in front of the house we would certainly make, if you like, a few beautiful Welsh ponies out in the pastures to keep the grass tidy, but be it ever so right in the eye of the sun, Mr. Powell, Dolgwili definitely was not for basic slag, for tractors, for awful old corrugated iron shacks and cows and sheep and mud and death and blood and guts and worry and mud and mud and toil and loss. No, Mr. Powell. Chicken-farmers, in a laughable sort of heartbreaking way, already we had been, Mr. Powell, you see.

And what, in any case, did we know of sheep?

Yet, right from the start, sheep pressures had insidiously threatened Dolgwili. It was almost as if sheep had a promo-tional campaign of their own, as if the Dolgwili slopes had been ordained for sheep in some undeniable natural order, as if to refuse them access there was to sin against some holy writ of nature. Conversely it came to seem an absurdity, even a wrong – as well as an impossibility – to make a garden of the pleasant slope of sheep pasture by the cottage. Welshmen stared and shook their heads in doubt at the queer East Anglian notion of planting prunus subhirtella autumnalis, John Downie crab, juniper and yew and lilies and delphinium and verbascum among mounds of heather and multi-coloured shrubs where for generations only sheep and a few sad cows had walked.

"Sheep, now . . . " old John Jones had mused a year or more back as he had leaned on the rusty gate and watched his grey ponies now safely back grazing our grass after that first expen-sive escapade into my new garden. "Aye, sheep," he had sug-gested unsurely, dropping the words into the suggestion box of my mind whilst we stood watching his inelegant grey ponies. Their inelegance had just been enhanced by their digestion during the previous night of all the young branches of my newly-planted prunus Tai-Haku, my robinia Frisia, a weeping birch, salix matsudana with half a dozen shrub roses for dessert.

The good John Jones had come running at our telephoned squeals of anguish that morning, trundling round to Dolgwili immediately with his old van loaded with barbed wire, stakes, choppers, hammers and doubtless a complete set of old iron bedsteads which, on the evidence left about Dolgwili's hedges, seemed to be the standard equipment for hole-filling hereabouts. Old John Jones had also brought an even more venerable white-haired character whom he claimed was his uncle and together they set about rescuing my garden-beginnings from his ugly brutes.

John and Uncle Dai had yelled and waved with great authority at the invasive ponies, which snorted and stuck their quick hooves savagely into my infant lawn, née sheep-pasture, as they careered back through the gate to their correct place in the field. Yards and yards of wretched barbed wire, gingerly handled and inexpertly fixed, sealed the gap quickly enough, but both old men smiled a lot, gently and nicely, and addressed me as 'sir' until it was obvious that they shared the viewpoint of their ponies – that gardening here was an insanity.

So, the situation restored – except for my beheaded poor young trees – John Jones and Uncle Dai leaned on the gate and murmured,

"Sheep now . . . Suit you fine, sheep would. Yes, yes. A few Speckleface, see. Put a Suffolk tup on them. Yes, yes." (John Jones was a great 'Yes, yes' man.) They stared ahead of them a little wistfully, right through their clumsy grey ponies, towards a dream of springtime lambs. "Do well here, sheep. Yes, yes."

John had been the last, and most sensible, answer to our advertisement offering our grass at 'tack.' The first answer had been another Jones, a younger man so pleasant that he could never bring himself to allow the awful details of how much money he was prepared to offer to cross his lips. All one afternoon we sat and sipped tea whilst we politely spoke of the weather, the season, how many sheep he had, his car, where he lived, his car, the weather, his car, and every time I was insensitive enough to approach the subject of rent, Mr. Jones would fidget uncomfortably and get back to weather or car until it seemed that we might have to offer him a bed and perhaps negotiate further next morning. Quite by chance, however, the name of the agent through whom we had bought Dolgwili occurred.

"Ah," my Mr. Jones exclaimed in delight. "You know Mr. Owen? He is my agent, too."

And he went on to chatter of our mutual Mr. Owen with such acclaim that, late in the day, I saw the light and said, "Would you like me to go and see Mr. Owen about this matter?"

Mr. Jones nearly fell about in gratitude, shook my hand with savage relief, talked a little more about weather and car and departed. For ever. Mr. Owen wanted to know which Mr. Jones, and I did not know, and anyway the rent was only peanuts.

The second answer was from one who claimed his name to be 'Gitimis'. He had phoned from twenty miles away about the tack, and the name, here in Thomas-Williams-Jones country, seemed quite odd. Gitimis arrived to inspect the pasture on a hot August day. He wore standard tack-inspecting kit – an old raincoat, wellingtons and thumbstick. It was very hot. Whilst we sunned ourselves in front of the cottage, Gitimis limped all over the farm. When he returned he still wore his standard kit and he ran with sweat. Unlike the sensitive Mr. Jones, Gitimis came straight to the point and offered us more money and better terms than ever we had dreamed of. He would fertilise, hedge, mend, accept all responsibility, and pay well. We thanked him sincerely and he limped back into the hills around Llanybyther and we never saw him or heard from him again.

"Damned funny name," I pondered.

"G. Thomas?" my wife, Phyl, suggested.

Ah . . .

We could never understand each other, them and us, at first. Except old John Jones.

"Why do you keep ponies, anyway?" I had asked him. He never sold one, or rode one, or broke one, or admired one. All six greys were the haphazard produce from one old grey mare which John had once accepted as payment for a bad debt. In his trade he always had need of accommodation land – he was a market dealer – and thus it was that his ponies had come to graze Dolgwili pasture – and garden trees.

"Why do I keep ponies?" He had smiled in slight embarrassment; perhaps he had never asked himself before. His thick red fingers had removed his cap and then scratched at his thin hair. He hit the gate with his stick, laughing.

"I'm blessed if I know, you see. Yes, yes. But sheep . . . I can bring you some sheep, you see, yes, yes."

"But I don't want sheep. I just want a garden."

"Ooooh, there we are then. Yes, yes."

Sheep, sheep, sheep, everybody said sheep.

"Ah, yes, ponies lean, you see," Hannah Rees had proclaimed with doubt, and a phrase to remember. "Ponies lean. Lean on posts and gates. Aye. Sheep would be better. Do well here, sheep."

She was passing through Dolgwili on the way up to visit her brother at the farm above and had noticed my beheaded garden trees.

"There is something else. I wonder if you would be kind enough to ask your brother could he bring his tractor and trailer down to transport our coal up from the road, would you?"

Twice we had phoned Mr. Thomas before. Each time he obviously had not understood.

"Yeess?" he would softly say, and I would change the wording of my request, would speak more slowly and distinctly, but again "Yeess?" was all the answer.

Now, similar puzzlement attacked Hannah's face. Repetition of the request only produced repetition of Hannah's puzzlement.

By now Hannah's husband had arrived, a small smiling man in wellingtons who always walked what seemed to be an obligatory forty yards behind his bustling wife. We did it all again.

"Coal?" Hannah asked him. "What's coal?"

"Coal," he repeated in that soft Welsh voice which so shamed my loudness. "Glo," he translated.

"Ah, glo," Hannah burst out in pleasure. "Yes, of course. Glo. Yes, I'm sure he will. Tomorrow, if the river is low enough. Aye. Sheep would do well here. Always used to be sheep at Dolgwili." And she passed on, and, of course, her excellent brother came and moved our coal for us, and he, too, nodded and said, "Aye, sheep . . . "

Everyone said sheep. Undeniably the acres tramped by John Jones' grey and inelegant ponies gave us little pleasure. Thistle and nettle proliferated, the ponies had ruined my first plantings

16

and now John Jones had doubts about the value of the tack. Everyone said sheep, and of course, even back in our old East Anglian days a lamb on the lawn had been a seductive little dream.

"But we don't know the first thing about sheep. What do you know about sheep?" I challenged Phyl.

"Only that they have lambs." Witheringly, as if 'that is all ye know on earth and all ye need to know'.

So now, right at the top of our steepest ground, the official man with all the answers was looking out across the green patchwork of Dyfed towards the Atlantic. He strolled about the ragged hedges and thin soil, composing, surely, his report that must confirm there could be no future on these hills even for men born to them. Bracken, gorse and hawthorn were creeping out from the woodland to lay thorny hands on pastures where John Jones' grey ponies munched out their term. The hotch-potch woodland itself was a nightmare of bramble and black-thorn, of fallen birch and young ash. Dolgwili's access was terrible: across a river by ford, impassable for much of the winter, and then over a railway crossing, albeit the railway suffered only one slow and doomed train per day; our acreage was laughably small and our buildings, like our experience, almost nil. How could this expert come down now and possibly recommend sheep-farming to us?

He was still smiling gently when he rejoined us, and as we sat at coffee it became obvious that he, too, was in the pay of sheep. He enthused, damn him, about Dolgwili's suitability for sheep, he praised with his beautiful Welsh voice our south-facing situation and natural drainage, said how enviable we were to be able to run a small flock of, say, Welsh Halfbreds on such a place. What could be better?

Start off with a couple of dozen three or four year olds, ewes that knew what it was all about, Welsh Halfbreds or Beulah Speckleface. Local sheep, not yearlings. Put a Suffolk tup on them, not too early, for we did not want them lambing in January. Or a Border Leicester tup if we liked. They would all look after themselves pretty well. Lamb out on the hill, you see, and the lambs get up very quick. Just to keep an eye on them, and we could reckon to make ten quid an ewe. Just feed them well before lambing and there would be no trouble. Say you get ten to fifteen quid per ewe and you lamb at 150 per cent, ram depreciation would be two quid at most, a bit of concentrate, vet bills, odds and ends about three quid, ewe depreciation about two quid. Oh yes, there'd be a profit for sure. Yes, very nice indeed.

"But really, I don't know the first thing about sheep – " I had begun to protest, but he smiled disarmingly, waved an irresponsible hand and claimed that it would be a great advantage.

"So much the better. You'll be all the more willing to take advice from us," he said, taking his leave. He cared nothing for my dreams but waved farewell from the bottom as he climbed in his car, while Phyl waved back and smiled and smiled in triumph.

No, but . . . Come back. That's not the idea at all. Had he no conception of my Country Gent status? Did he not understand how besotted with woodpeckers and lilies I was becoming? How could the saluki chase across the splendid hills if they were to be cluttered up with blasted sheep?

3   A different country

For us, Wales was a liberation. Suddenly anything was possible. Old home-town inhibitions sloughed from us as we abandoned our native confining niche to land on our feet in all that space, right in the eye of the sun. Salmon-fishing, rhododendrons, aloneness, even sheep-farming, dozens of formerly unthinkable leaps in the dark suddenly were in our mind. So much space, so much freedom, made it a heady experience. This was the ultimate extension of a rural dream which had started on a third of an East Anglian acre with a dozen mallard ducklings in the garden pond, a donkey in a handkerchief-sized paddock, bantams in the garage, nearly a lamb on the lawn and the threat of bats in the belfrey. All that had burst at the seams eventually. We had graduated then to ten Suffolk acres of smallholding with deep-litter hens, goats and Shetland ponies, and so on at last to this whole twenty-three acres. What was not possible now?

Ah, but what *was* possible now? First thoughts sobered as we began to experience Wales-as-she-was and not Wales-as-we-thought-she-was. It was rougher, harder country. The hills were steeper, the rivers faster, the air soggier and cleaner, the days longer, the people friendlier than anything we had known. All was strange. It was a different country. This was Welsh Wales, and although it was extremely foolish of us not to have known it was a different country, it is a foolishness the English seem prone to. Builders, neighbours, auctioneers, shopkeepers, all spoke Welsh naturally; English came with difficulty. Our own attempt at Welsh foundered completely within a month, for the motivation to learn was too small and the difficulties for our elderly tongues too intricate. Isolated linguistically as well as topographically, we settled on our hillside to a slower, more old-fashioned, rural way of living. It was, then, an intoxication.

The valley was warm and moist. Trees flourished. Grass flourished. Sun and rainstorm flourished. Hordes of slugs flourished, and wildflowers, rabbits, foxes, butterflies, house-flies, horseflies, wasps, midges, ants, mice, owls, tadpoles, all things flourished. Even frost. Right in the eye of the sun by day, by night Dolgwili slipped down into the moon's cold pocket. The cold and ever-singing Gwili poured through a chasm at the bottom of the acre-plus which was meant to become my garden, making of it a frost pocket. Frosts still came very late in spring, even into June, and left us free only until September. The great winter rains collected on the hills all around and soaked down to the river, keeping our soil cold until well into May. Phosphate and potassium deficiencies were vast, so that seedlings planted out in June remained seedlings for the rest of their lives – which rabbits, mice and slugs decreed should be short anyway.

All these things we were to learn almost imperceptibly. We were like children in a forest; it was all strangely new and wonderful, and perhaps would turn out to be fearful. We had to keep our eyes skinned. Sometimes I imagined that I could hear trees being torn down in the woods, and that in a minute I would see a dinosaur emerge across there on the hillside. I even eyed the grass where I was to garden as much with suspicion as with pleasurable anticipation; could I really make a garden here?

Could we really manage sheep here? Well, that was what the man said . . .

4 Sheep hustled in, sheep hustled out

The clear-out sale at Llynwalter, Alltwalis, started promptly at ten-thirty on Saturday, 15th October, and by then it amazed us to find quite three hundred cars, vans and lorries already untidily parked in the meadow near the farmyard. Seventy-four head of cattle, six hundred and nine sheep, thirty-one pigs, all the machinery and implements, some household furniture and '2,000 bales of Meadow Hay in Good Harvest' had to be sold all in that one day, so the silence and stillness with which the crowd heard the auctioneer introduce the sale from his mobile rostrum – with warm and kindly reference to Mr. and Mrs. Evans who were now retiring from farming – that silence and stillness were to be the only such for the rest of the day. The first huge Friesian with calf at foot, immaculate in a well-groomed black-and-white hide, was ushered before the auctioneer to fetch £245 and the calf, separately, fifty-eight pounds. Thereafter all was to become bustle and noise as the rostrum with its escorting crowd of buyers moved about the farm, whilst we retired to sip coffee and to await, shaking gently with excitement, our Mr. Thomas and our subsequent purchase of sheep.

Across the sunlit valley, autumn had already seared a plantation of larch to russet brilliance, and the white dots of the eternal sheep confettied the black-hedged square pastures, spread like a green knitted-patchwork blanket across the smooth landscape until it folded and dipped into difficulty and dingle. There seemed a sense of destiny all across those fields. Nobody else knew or cared, but somehow we had arrived on the edge of sheep-farming ourselves.

We wandered about the farm, anxious for our Mr. Thomas, looking in at sows about three yards long, comatose and faintly mysterious in clean straw, at impatient calves, at bits of machinery whose function was beyond our imagination, at great tractors of awesome power, and on inevitably to the sheep pens. Everywhere Welsh farmers trudged in heavy boots and good cloth, thick cane walking-sticks hooked on their arm, often hauling a farmwife in tow, functionally dressed, beyond the claw of fashion. They stood inspecting sheep, eyes keen, hopping over into a pen to feel the back of an ewe they fancied, casting comments over their shoulder in Welsh and thereby robbing us of knowledge.

"They look nice," Phyl ventured bravely, drifting to a stop beside a pen where we stood, nervous and shy, discussing whether they were rams or not, a tiny island of English ignorance stuck in a flowing tide of Welsh knowingness.

Our Mr. Thomas – recommended to us by Dai Petrol as the best breeder of Speckleface in the district but not, as it turned out, having stock now for sale – was to recognise us by the huge green and red scarf that I wore. Nobody there could have felt more self-consciously alien, for although a biting wind savaged the slopes where we stood, Welsh farmers simply did not wear scarves. True, each face that we expectantly scanned as we awaited the challenge of our Mr. Thomas was inclined to red-and-greenness in that bitter wind, but we saw too that this was something more than a mere clear-out sale. It was a true social occasion. Friends and neighbours mingled with avid interest to inspect and honour Mr. Evans' lifework of breeding livestock. This was Saturday afternoon. Elsewhere thousands would be cheering their sporting heroes, but here the recreation was to stand and chat at the sale ring, to consider rams and tractors and perchance to make a quick quid by a crafty resale. They might buy, they might not, but either way they saluted the retirement of a worthy farmer by their presence, or so it seemed to us newcomers.

The patient sheep stood penned in tens, in two long lines of hurdles. When our Mr. Thomas – quite properly half-an-hour late according to protocol – shyly found my scarf, he turned out to be so softly and so very quickly spoken as to be almost incomprehensible. Together we drifted along the pens in a pre-sale inspection of some bewilderment, for in turn he seemed as little able to understand us. He was small and wizened. The biting wind decorated his bony nose with a permanent dewdrop, or icicle, not dislodged even when he dived into a pen to grab an ewe, to turn it on its back, to examine its teeth, its udder, its feet. He seemed to be confirming that this was good stock, so in a semi-articulated and gesturing discussion we tried to commission him to buy us about twenty ewes, three year olds and sound. The dewdrop-icicle remained on his nose all day, and we entered him in our family mythology as Thomas Dewdrop, alongside Thomas Panteg, Thomas Uptop, Thomas the Post, Thomas the Pen, Thomas Etcetera.

The auctioneer's rostrum, Land-Rover drawn, had swept in a whirlwind of selling across the farm, sucking its throng of buyers and onlookers along with it through the churned mud and over the bruised grass in cheerful pandemonium. Too suddenly it was at the sheep, taking us by surprise, sucking us too into the crush of buyers.

The first ten ewe-lambs – too young for us – were hustled in by drovers to the hurdled sale ring. The sweet innocents stood there in the straw suddenly awed by the rings of staring, shouting humans. Before the bidding was completed, drovers hustled them out and other drovers hustled the second ten in and then hustled them out and they hustled and they hustled and the bids flew over them like bullets, whilst the ceaseless drone of the auctioneer's chant and the endless movement of his bid-hungry eyes produced a sort of hypnotic fascination. Poor Thomas Dewdrop had begun to look distinctly ill-at-ease. When lots seventeen and eighteen – our chosen ones – came in, his winking eye worked too slowly, or his head nodded too indefinitely, so that his bids got lost, thereby spreading his unease to me. Being the best Speckleface breeder around obviously did not make him the best bidder. Phyl had withdrawn fearfully from the initial crush, so when with difficulty I turned to mouth my concern to her on the outskirts of pandemonium and would it not be better for me to take over the bidding, she merely smiled her lovely smile and nodded vaguely in a quite serene sort of way. The bidding was so hot, Thomas so Welsh and myself so ridiculously English and anxious that total confusion fogged our minds. At this rate we should buy only sweet Fanny Adams.

I pushed through bodies and plucked Mr. Thomas' sleeve. He was well wedged in and his dewdrop had grown a bit more, but he seemed to mouth back over his shoulder at me that

"Four year olds are best for you," which was just as well, because already that was all that was left.

Sheep hustled in, sheep hustled out, the auctioneer chanted, his assistants poked and pointed their sticks at bids, buyers winked and nodded, women laughed, and still poor Thomas had bought nothing. A second long-distance consultation by backward glance and mouthing to Phyl brought a decision to do our own bidding, a decision almost immediately abandoned as we discovered we were bidding against our own man. I had just outbid Thomas and had to take to sudden head-shaking to reverse the bid, but a stick-pointing assistant at the auctioneer's elbow, an ugly accusing man, shouted at me and kept pointing his damned stick straight down my throat whilst he displayed a great and public contempt for me and my daft scarf. It was bedlam. It was certainly no fit way to mark our entry into sheep-farming, and I slunk away in despair just as poor Thomas half-looked towards me at last to claim something with a positive nod of his head.

"Did you get that lot?" I bellowed in near disbelief.

"Lot thirty one. Sixteen pounds," perhaps he said.

"Get another, quick," I mouthed, me who was against sheep-farming.

He nodded to me and thereby, probably, bought the very last pen, six rather scruffy elderly ewes with dark fleeces at fourteen pounds each.

Whew.

The ring broke up, the auctioneer and his rostrum went bowling off attached to the Land-Rover, a smiling wife rejoined us to squeeze my arm in delight and we trooped off with our chastened Thomas to find the office.

In the little yellow front room – how modestly they lived – behind the clerks, the farmer's wife and neighbour friends dispensed on-the-house whisky, port and ale to the jolly throng – it really was – of chequebook-brandishing buyers waiting to settle. Perhaps the farmer's wife did look a little sad, but it was so nice in there. Our faces suddenly burned, not only from the bitter wind and burnishing sun we had endured all day, not only from the lovely port, but, fools that we were, from a different brand of happiness. Phyl's face was radiant.

I raised my glass and whispered, "To us. Sheep-farmers."

Then, as our own chequebook turn came, our Mr. Thomas discovered that four lots, not two, had been knocked down to him, thirty-six ewes, not sixteen. No matter, he laughed ruefully, as he finally accepted our fiver with dying protest and surprised thanks, no matter, for they were good ewes. He would keep the extra ones for himself. They would be no ill store, and thank you very much, yes, and he'd come over and give us a hand at shearing.

What, us? Shearing?

5 Thou fool

Thomas Dewdrop had gone. It was late afternoon; dusk threatened. The sudden realisation that we were now totally responsible for sixteen real live ewes broke over me as soberingly as a bucket of cold water.

"So how do we get them home?"

Three large livestock transporters, presumably loaded, were just trundling away down the farm road. Self-sufficient little Morris vans were hurriedly backing into loading positions by the sheep pens. Land-Rovers were carting trailers about busily.

"Dai Petrol warned there might be a shortage of transport, being Saturday."

"Perhaps old John Jones is here somewhere with his van."

"And Dai Petrol warned, didn't he, not to leave our sheep unattended."

Our sheep! Thou fool.

We chased about the field forlornly seeking transport and began to feel lonely and foolish. This, I had always known, was how it would be henceforth.

"Didn't Simon offer to help us get them back?"

It was the end of the fishing season, or, if you like, the end of the world, for our son. This morning, as we left for the sale, he had been desultorily cleaning out his Renault Four, a twin to our own, which acted as his kennels, gunroom, tackle room and lumber room. 'Cleaning out' was a process of mooning lovingly over his spinners, guns, reels, rods, terriers and spaniel, of examining any old mink or rabbit pelts that he was drying, of collecting up any odd feathers put by for fly-tying. The spontaneous offer of help he had made was almost unbelievable, for he had never troubled to hide, as an agricultural trainee about to start his second-year course, his derisive disapproval of Phyl's daft sheep notions – or, come to that, of all parental notions. By vocation he was a poacher/rough-shooter. He wanted Dolgwili to stay a wilderness wherein he could hunt and fish; he wanted no truck with sheep. He did not actually object to giving over a few hours a day eventually to the earning of cartridge-money, rod-money, dog-money in what had, inside the household, come to be called Proper Farming, but in his usual few words he had plainly ridiculed any possibility of us farming Dolgwili. Nevertheless, he *had* offered. Come to think of it, he it was who had suggested the economy of bringing them home ourselves.

So, Phyl would stand guard over her precious ewes, I would rush home for Simon. Each car would then carry four sheep, twice; our sheep-farming's start would be marked not only with a rather appealing economy but with the involvement of our son. Good.

But his life had changed again. A girl had come into his life since this morning. He frowned; would we complete the transportation by half-past six when he was to meet Liza?

And by God, his subsequent driving of that poor little Quatrelle from Dolgwili to Llynwalter, his throwing of our brand-new old sheep into both our cars and then the return trip, all to be duplicated, proved frighteningly that he had no intention of not completing the job by six-thirty. He drove and moved with a purpose which, both for car and boy, we would previously have deemed impossible and which could only possibly be motivated by love – and that not for ourselves. Troubled and breathless, we began the first shepherding of our lives, mother and son at the car end, father at the Dolgwili end. As we drove the last few ewes through the farm gate, I saw that one was limping very badly. The leg was obviously broken.

"How did that happen?"

She shrugged. "Carelessness, hurry. The whole lot nearly got out on to the road. 'Ach, they'll be all right. No need to worry,' he kept saying. That one caught its foot in the gap between the bumper and car body as he unloaded them."

The ewes were already settling down to graze. We watched them silently, oppressed by our anger at the broken leg.

Simon came hurrying down from the house. Doubtless it was six-twenty-nine. Probably he was still wiping after-shave lotion across his cheeks to cover the ewe smell.

"Can you manage by yourselves?" he asked and already a touch of contrition softened his tone.

"We'll manage," I said grimly, looking into his face so that he should see what I thought of him as a future stockman, and he disappeared, trotting off to his damned Liza and the demands of his loins.

We shut our farm gate on our new flock. They wanted nothing to do with us and I did not blame them. They began to climb deliberately, the last one carrying her leg off the ground, and went straight up to the highest part of the farm.

I had been right; sheep were a terrible mistake. One broken leg, I kept thinking, just one unnecessary broken leg and everything was muddle. Tomorrow we would need a vet. We would have to catch the ewe. How? Where to hold it? Visions of old Dai Sheep and his grand-daughter tormented me already. We had no fences, no barns, no pens, no dogs, no lassoo, no shepherd, no hope. The sixteen ewes were fast and loose on all of Dolgwili's twenty-three acres.

"Well? How does it feel to be a sheep-farmer?" I enquired with heavy heart as we prepared for sleep.

"That man from the ministry," she said, musingly.

"Didn't say anything about broken legs, did he?"

"Perhaps he did not realise that they could break their legs. He was a desk man, wasn't he?"

"Try and get a good night's sleep, my love. Tomorrow will be a hard day. In fact they all will be from now on."

Perhaps Liza had let him down, perhaps contrition had grown overnight, but anyway by breakfast Simon had called in a shooting friend named Caleb to make a pen from larch stakes and odd bits of chainlink in one corner of the pasture, already named Sheepdip from a rusting old collier's cart left there, derelict now but obviously once the farmer's sheepdip. His conscience satisfied by this construction – as indeed was I, for the stakes were well driven in – Simon then reached for his gun, whistled his dogs and departed. He was clearly determined to maintain his policy of No Truck With Sheep: we were clearly determined to manage without him.

Our flock sulked right up on top of the hill by our boundary hedge. The plan of campaign for this our first nervous brush with sheep entailed starting by calling up our reserves – daughter Trudi and her husband Peter had lately joined us at Dolgwili – to form a thin red line which would advance by scaling the ridge and then slowly sweep round to drive them down into Simon's pen, where clever us would guard them until the vet arrived.

Halfway up the hill Trudi, who from nursery days had never ceased looking over her shoulder for bogeymen and who thus saw everything that moved, called down, "There's a red car just drawn up, Dad, and a man in a white overall is getting out."

Lateness and delay had become the order of our days; not for one moment could we have anticipated so prompt a vet. I hurried down to the gate, apologised embarrassedly and explained. He said he would call on his return trip, in about an hour and a half.

Up the hill, pausing long for breath as we climbed, went my sheepdog-family and my apprehensive sheepfool self. Mad we were. The sheep, of course, expecting more broken legs no doubt, began to scatter. It was new ground to them as well as to us and there was as little understanding between us all as there had been with Thomas Dewdrop yesterday. Over and over again the four of us clambered up and down and across that damned steep hill and each time the scared flock would break up and make fools of us as they ran, with always the broken-legged ewe plucking at our sensitive heartstrings, Trudi's in particular. When I paused for breath and patience, it would not have surprised me to see, far below in that little meadow down beside the river, poor old Dai Sheep and his Bo-Peep watching and laughing in turn at our mad scurrying, just as all those years ago we had watched them.

"We'll never do it, you know," I addressed our council-of-war there on the hopeless hillside.

We stood and considered.

"How about phoning Uncle Dai?"

When old John Jones had come with his Uncle Dai on the morning after his ponies had chewed up my garden, Uncle Dai had brought with him his sheepdog. Together, it seemed, they did part-time shepherding for John, and together, too, we had since discovered, they were the Royal Mail.

In the happier days before sheep, a man had astonished me

one morning by appearing over the skyline, climbing over the hedge and coming down the Steep to the cottage with a letter. Uncle Dai. He had walked, with his thumbstick and Juno, three miles from the village post office to get a signature for what turned out to be a licence-reminder or some such, and then, after a chat during which we tried to sort out the unsortable mail situation – for we seemed to have three different addresses – he walked three miles back. Since then we had established a typically Welsh solution to our mail delivery, which entailed adopting a fourth and unofficial address, whereby Uncle Dai was saved all that trudging, so it could just be that we had a pull there.

A pull was hardly necessary. Of course he would come, he said, just as soon as we could fetch him.

Uncle Dai was white-haired. He wore good sound boots and the thick old black trousers with red piping which postmen used to wear. His border collie was most enviable, and whilst my womenfolk revived themselves in chairs with drinks, I took him to our makeshift pen and showed him our frightened flock cowering at the top of the hill.

I blush still. I blush with shame as I remember how terribly arrogant the English so often are.

"If," I actually said to Uncle Dai, "if you can push them along that top hedge to the corner and then bring them down the hill along that other hedge to the railway fence over there, and then along . . . See?" Pointing, explaining. "And so into the pen."

Could that really have been me?

He was a fine-looking man, even in age. His features were good – strong nose, straight bones – but beyond that it seemed that he had the additional beauty bestowed when such a man devotes himself to the outdoors and a concern for his animals. When I had the green English temerity to issue advice to a true Welshman on his own Welsh hills with his own splendid Welsh dog at his feet on just how to go about shepherding Welsh sheep, just the merest hint of amusement touched his lips and creased his eyes. But he said nothing, whistled Juno and together they started walking.

He walked directly up that steep hill without zigzag or halt for breath. He turned back near the top and, as he descended, our flock was a solid phalanx at his heel, with Juno behind them in total control. He brought them straight down and he penned them without the slightest difficulty. It had taken perhaps ten minutes. He must surely have showed some breathlessness, but the way Uncle Dai stands in my memory is serene and masterful, a simple straightforward man with a lovely black-and-white friend at his solid feet, an admirable pair.

It was nothing, he said, brushing aside our bubbling thanks and in a minute the vet arrived, a swashbuckling man in deep waders, bloodstained overalls and wide moustachios. Together they pounced on the injured ewe, tied its three good legs together and applied a hot plaster bandage to the broken one. Our only jobs were to fetch water and to hold the treated leg for twenty minutes while the plaster set and while the vet and Uncle Dai went through the flock one by one, worming them, turning them on their backs and treating their feet with a fearsome knife for foot-rot. And all about them was chaos. Simon's pen had disintegrated at the first push of the frightened flock and without Juno the whole lot would have scattered. The chainlink concertinaed on itself and tangled, the large stakes quickly dragged out of the damp ground so that sheep and vet and Uncle Dai and we drifted this way and that in a loose maul contained only by the frisking Juno. The two men worked without grumble, bent and wrestling with the nervous ewes until the work was done. The ewes shook themselves and walked quietly off to graze. The vet and Uncle Dai, with the quiet 'and cheerful words which we had come to see as the hallmark of Welshness, did what they could to ease my wounded feelings as they stood on the bruised grass packing away the worming gun, the knife, the all-purpling aerosol. And when I took Uncle Dai and Juno home, he would not take a penny for all his trouble.

But we were shamed by the episode; it had been a shambles of sheep.

6   A broken-down old gentleman in black shoes

Came Sam-Ram.

Down on the road John Jones' all-purpose van pulled in, and within minutes the good man himself was leading towards us on a halter a large black-faced ram, who even at that first sight and even to us gave just the slightest impression that he had seen better days. Yet he also gave the impression of such masculinity that it made more ridiculous our inability at that Llynwalter clear-out sale to tell rams from ewes. He had a noble head, but, alas, he was no longer young. His fleece was respectable, yet somehow creased and dingy – like a man wearing an old mac – and his feet already gave the impression that he was wearing broken-down old black shoes. Nevertheless, he was *our* ram, and thus suddenly transmuted from being just one more cast-off ram in a Welsh marketplace to being something rather special.

As the ever-smiling John freed the ram from his halter, Phyl named him Sam-Ram, being at that time gripped by a strange but fixed notion that only the name Sam truly conveyed masculinity. We had Sam-Dog, Sam-Cat and a dirty old neighbour, Sam-James, so that Sam-Ram's arrival brought conversations more than somewhat apt to become Mrs. Feather-ish.

It was late October. The ewes had proved to be obstinately stand-offish, keeping to the top of the Steep day after day. The ewe with the broken leg was usually solitary, moving only enough to maintain a token contact with the flock and that still on three legs only. When Sam-Ram arrived, he baa-ed a bit from down on Sheepdip to his sixteen wives up there, tasted a little of our grass and then promptly clambered straight up the hill and proved his Sam-ness by shamelessly tupping one before our very eyes.

"Yes, yes," old John laughed and said, his red face beaming with satisfaction, "there we are then," as if to underline how settled things now were for us. Tup them with a Suffolk ram, the man had advised in that idealised blueprint, lamb them down in March, and here we were already counting on our fingers the weeks between now and spring lambs.

"How much do we owe you, then, Mr. Jones?"

"Oh, well now, I had to pay rather a lot now, you see. Yes, yes. Not too many at the mart today, rams, you see. Yes, yes."

"That's all right. I understand. But how much?"

"Well," reluctantly. "Twenty-five. But now I tell you. I'll come along later and we'll settle it all up some time. Write it all down I will."

"All right. As you wish. I'm very grateful to you, though."

We had made outrageous use of John Jones, getting him to open all sorts of gates into this Welsh heartland where so accidentally we had come to live. He was never to charge for half the things he did for us, for whilst we assumed our relationship to be a commercial one – he was, after all, a market-dealer – he himself seemed to assume it was a neighbourly one.

His grey ponies had gone now, spirited away via a country backway from Dolgwili across other farms of which we knew nothing. Those ponies had not been halter-broken and would never come to hand for us, so how John managed we never knew. They were gone, and soon in their place came our own Fleur, Syndod and Marigold.

Sheep, they told us, did not like long grass. Sheep need ponies or cows to graze alongside and for this reason, for their beauty and for company, we bought two pedigree Welsh ponies and a donkey. According to the pastoral pastiche composed by our ministry man, all we now had to do was to leave it to them. But as I stood among my new plantings visualising further plant-borders, I knew all too well it would never be like that.

For it had never been like that. We could look back down the pathway of our animal-years, long before Dolgwili, and see it strewn with plucked feathers, still bodies and sadness.

The marvellous yellow-fluffed ducklings Simon had reared long ago had grown to messy mallard with a secret passion for our neighbour's vegetable patch three doors up. As soon as we had departed for business or school, they would depart up the road and devour poor old Mr. Turner's best cauliflowers before returning to sit with innocent faces to greet our own return. They had had to go. Our silkie bantams always had unaccountably turned their feet up, not one pheasant poult had survived our summer care, foxes had dug into our pullet acres to perform massacre, our goats had either half-strangled themselves when tethered or else tunnelled out of any compound to head directly for my garden, to annihilate my favourite rose bushes. Our Shetland pony had broken into the deep-litter henhouse one night, gorged itself on layers pellets, to lie there so terribly dead next day, leaving me weeping and shattered. Dogs, too. Boxers whose puppies came white and were therefore doomed; the dalmatian who constantly suffered false pregnancies; the accident-prone saluki; the boxer pup whose little contorted body we nursed for two years through heartbreak and worry. Heartbreak and anxiety had been our companions with all our animals. Even the splendid Tiger Lil, Simon's Jack Russell terrier, although throughout her joyous life she had been the fiercest, best and most tireless hunter, although she had successfully, even easily, raised all twelve puppies of her record litter, and although by our fireside she was the gentlest and most affectionate of pets, nevertheless she too brought us to tears in the end.

Like Mother used to say, It always ends in tears.

Tiger Lil disliked, above all things, diesel railway trains and in particular she disliked the slow old milk train which trundled and squeaked its way up the valley past Dolgwili each morning and back each afternoon. One morning in those early Dolgwili days when we were still fighting the good fight with builders, we had preoccupied ourselves with wallpapering whilst Tiger Lil and her son Urchin sported outside. Later Urchin came home alone and it is absurd to claim that he had tragedy written all over his face. Nevertheless he told us that something had happened. I went out searching.

Tiger Lil's dislike for the diesel train had been too much for her perhaps that day. She had, I felt sure, attacked it. Weeping buckets, I collected her shattered remains to reassemble them bit by bit into a small grave inside our farm gate.

So we had had our fill of animal-tears, and from that had come my reluctance to farm, but rather a desire to fill my hours with the gentle and tearless keeping of flowers and fruit.

Yet how could one live without animals? To farm, to be accompanied by splendid creatures in a green country, this had been the pastoral dream ever since that hard bed in Room Eight, Block Six of Stalagluft One. To return to Suffolk, to live a quiet life with a lovely wife in an old countryside of thatched cottages and water-meadows, nothing more than that did the heart desire. Perhaps I did not desire it enough. Perhaps I desired it too much. For one reason or another it never quite came true in Suffolk, but now I had come again to the very threshold of the dream-come-true at Dolgwili. I could not perceive that any life could be more desirable than to live simply in green space graced by the elegance and interest of woodpeckers, roses, trout, apple trees, mountain ponies, dogs and, if it had to be, sheep.

So daily on Sheepdip my binoculars would check over the kingdom I ruled, revealing Sam-Ram – to whom of course we were already affectionately bonded by our common Suffolk ancestry – gradually persuading his shy Welsh fools to come off the shelf up there and investigate the flatter parts of the farm, the alternative grasses, to rotate and to become more matey. In truth the binoculars did not reveal our flock to look the most attractive creatures in the world. The flock at Llynwalter had been of excellent repute, but our Mr. Thomas had hardly bought the pick of them. They were Beulah Speckleface, a breed developed about fifty miles away over the last century, hardy yet manageable and well-suited, the books said, to our terrain. Several had uncharacteristic dark fleeces and some looked ragged, but worst of all was their feet; they were a load of walking wounded. One or other was always limping, and soon poor old Sam-Ram too was hobbling around in those broken down old black shoes of his. True, the vet, on that first encounter, had had to use his knife somewhat drastically on the flock's feet and they would still be recovering from this, but nevertheless, as time went on, we began to worry more and more about foot-rot, even though Sam-Ram seemed to be performing admirably.

By mid-winter a simple three-bay shelter had gone up on Sheepdip ready for the coming of lambs, foals and accident, together with a paddock in which to contain the flock at times of treatment. Furthermore, we had patched together a tentative arrangement with Simon whereby he would part-time shepherd the ewes and undertake the more daunting physical tasks. That the pattern of Simon's shepherding should turn out to be variable ought not to have surprised or disconcerted us, for he was, after all, only an apprentice-shepherd and that without a master-shepherd to guide him. So his announcement on one day that all the sheep looked fine and thrifty would be nullified on the morrow by his dire warning that quite a lot were limping and we really must get the flock in for treatment.

The drenching days of January soaked through into February; sheep, ponies and donkey lay abed late each morning up in the wood, loath to leave that comparative comfort to go out into the rough weather even for our waiting breakfast nuts. Similarly I was as loath to venture, turning my face from that miserable weather inward to the sowing of cineraria and geranium, to the fondling of fat lily bulbs before burying them in their pots of compost, to all the gentle pottering which the confines of the conservatory offered me. Only the everlasting need for change, however small, would then tempt me from the conservatory and the animals from the wood during perhaps some lull in the stormy weather, for us all to meet, to pass the dull time of day, to be counted and checked, before once again rain or dusk sent us all back to shelter.

One ewe in particular we began to notice these wintry days, one quite unable to cope and soon to be named Nellie Eighteen.

The original Nellie Eighteen of my childhood had been a poor feeble-minded woman who used to trudge round our town trailing a child and calling after a man called Boxer. Our ewe was similarly feeble-minded. Frequently, seeking winter leaves to eat, she entangled herself in the tentacles of bramble, charging this way and that in bleating frenzy to escape our anxious help as we approached, and thereby imprisoning herself the more. Her fleece was in tatters. Almost daily she rushed about the hill, bleating, searching for the flock, and if at roll-call one was missing, that, for sure, would be poor Nellie. Sam-Ram, Cripple, Nellie Eighteen, one by one each ewe began to emerge as an individual. Until then, sheep had been merely sheep,

36

but now they became Spotty, Darkie, Gentle, Black Collar, Scratcher, Greedy, and so on.

Morgan's Wood was fiercesome to middle-aged legs. From the river it sloped steeply, often precipitously, fifty almost useless acres of ash, hazel, oak and bramble. Trees clawed a desperate foothold among rocky outcrops and during winter gales many lost their hold and toppled over to a slow death. The slopes ran with water which here and there collected into marshy, peaty pools whilst rusted barbed-wire fencing marked some long past attempt to utilise this wilderness. Now it was strictly for foxes, buzzards, rabbits, owls and all the unseen life that flourished among the ivy, the mosses, ferns and brambles and also, now, our damned sheep.

Morgan's Wood was separated from Dolgwili by only a drainage ditch down which rainwater streamed all through the winter. The remains of some hedging had rotted on top of an accompanying bank, where also grew, at intervals, ash and hawthorn. It was not enough.

Almost daily our sheep broke through, seeking the thin woodland grass, munching blackberry leaves, leaving their shed wool on the wicked bramble thorns, and their shiny little black droppings as trail-signs in the mud for us to follow. The fear grew that one day this trail would lead – if we were too late to head them off – all the way across the wood to Morgan's farm itself where ours would mix with his. The urgency of this threat tired our legs even more, and, of course, it was trespass on to neighbour's land. Day after day it went on.

Fifty yards into the wood and about one third of the way up from the river, a natural drainage way channelled that winter's vast rainfall into a place where the conformation of rock was saucer-shaped. A messy sponge of forest detritus, moss and mud had formed, anathema to sheep.

That morning I had gone in alone; Phyl was already tired from yesterday's march. The sheep were, as usual, round the Point, a huge outcrop like a headland, which forced one either well up the slope or else right to the bottom by the river. The technique was to haul one's self from tree to tree, testing and then pulling one's weight along by branches, getting the wellingtons into trustworthy footholds, resting against tree trunks, cutting with secateurs any bramble which held woolfluff and

37

thereby cleaning up the trail ready for tomorrow, and thus eventually getting round to head off the flock and to drive it back home.

I drove them back to the Morass without difficulty. And there we stuck. They would not cross it. They would go up, they would go down, but they would not go across. They would not *all* go up, nor would they *all* go down. They split. If I herded the bottom lot, the top lot would turn back outwards and I would have to hurry up that damned slope again to head them off and then the bottom lot would have a go. They teased and teased and teased. Up and down, back and forward, we slogged it out, on and on.

I stood at stalemate and conceded that Simon, for once, had been totally right. In a state of mind-blown frustration I leant against an oak and reckoned we were short of about a hundred miles of fencing, a good sheepdog and a shepherd or two – and our senses. I was very tired, very hungry, very frustrated, for we were within sight of Dolgwili. I could see the house, the smoke curling blue from the chimney. I began to shout for Phyl. Here inside the wood it was quiet and still. Only the sheep moved, casually and without interest. Outside huge winds hurled low grey clouds from the Atlantic against the hills of Dyfed. My voice sounded silly; the wind consumed it. The infuriating sheep eyed me insolently each time I began a new manoeuvre, ambled tentatively off to try a new blackberry bush and wondered whether they should not try another dash off to Morgan's farm. I thought of all the sheep-days to come, of my lost garden hours, of lambing. I recalled the old Adrian Bell books about the shepherd's huts, the cold nights in the sheepfolds, foxes, lambing oils, sleeplessness.

You poor damned fool, I told myself, you don't want to spend your days like this. You want to make a garden. You knew all this would happen with sheep. They'll drive you crazy, mate. Worse; they'll kill you. Your poor stiff body will never be found, it will founder into the forest floor and only your rusty secateurs and your silly green wellies will remain to mark the spot where you gave your daft life for your stupid sheep.

I went home. I was totally exhausted. I told my tale bad-temperedly in the warm kitchen, I had a brandy and I went straight up to bed, where I slept soundly in my clothes for three solid hours in the middle of the day, whilst, of course, wife and daughter went to the wood and brought them straight home, chuckling to themselves, no doubt, about poor old Dad struggling by the Morass and quietly going barmy.

The answer, as it always is with animals, was food, of course. In future we must get them to come to us for food. The Llynwalter Speckles, however, evidently had never been fed; they simply did not recognise the new concentrates in my new trough as goodies at all. Fortunately they had among them a broken-down old gent in black shoes who, bless him, had evidently seen better days and good food. He straightaway finished off the trough single-handed and then came tamely and enthusiastically to hand for more, until the ardour of his feeding impressed even the Speckleface ewes. Little piles of nuts left strategically about the grass soon had them inspecting, sniffing, tasting and finally eagerly rushing about looking for more. Four or so still turned up their noses a bit, but most of them for the first time were now coming towards us instead of going away.

To this spectacular advance in management, we added the repair of the shelter roof and a promise from Simon to fence the whole boundary between us and Morgan's Wood. As though co-operating in approval of our progress, the great rains eased considerably. Soon it would be March.

On 5th March, the frost was particularly sharp. The morning was bright as I walked the boxer and saluki down for the mail. It seemed to my unreliable eyes, as we returned, that one ewe had detached itself from the flock right at the top of the Steep. Nellie Eighteen, I presumed. Yet something seemed to have been added. Something white and small.

"I have an idea," I entered the bacon-and-eggs kitchen and reported, "that lambing has started on Dolgwili. Right at the top of the Steep there is —"

She abandoned the frying-pan, threw on her dufflecoat, grabbed Grandfather Kirby's walking-stick and rushed out. Lambing indeed had begun.

7   A time of innocence

Never again, of course, would there be another lamb like the first lamb. It was like first love, a time of innocence, the everlasting miracle happening all over again. My solitary breakfast that morning was purposefully slow. The marvellous morning and the first lamb seemed, in the natural order of things, to belong to Phyl, and this fact the slowness of my breakfast acknowledged. Soon, however, armed with camera and the thumbstick, recently fashioned in hazel, which vaguely certified me as sheep-farmer-of-sorts, I clambered up the Steep to where Phyl stood purring in the crisp morning air over the blessed ewe and her perfect lamb. The thin grass was sparkling white with hoar frost, the lamb was strong and sucking, the ewe unconcernedly nibbling at the nuts with which Phyl rewarded her from time to time. All was right with the world. A single snapshot recorded the event for the family archives.

"Remember that? The very first lamb we ever had. Coo, I look quite young there. Seems a long time ago now."

We stood about high above the world just watching, saying the obvious things about how could mere grass and blackberry leaves ever be transmuted into such a fine little creature as this, about how on earth could life survive that sudden ejection from the womb's warmth on to the frozen earth, about could there be a twin still to come, what about the afterbirth, and was it a ram lamb or an ewe lamb, and so on.

The lambing had been quite immaculate. The sun came pushing up over the hill just as if it was an ordinary day, the frost began to dissipate and we to descend, touched with happiness. Today, at least, the ministry man was right after all; sheep-farming was all too easy and wonderful.

Morning by morning now, Phyl hurried out eagerly with her

bag of nuts and her stick, first down on to what was becoming our lawn to look up over our roof to the Steep, searching for more lambs, and then down through the gate into Sheepdip to call them to feed.

"Tot, tot, tot . . . Come on, then, my beauties. Tot, tot . . . Come on."

All the old timidity had gone now. Ewes broke into a gallop and in a moment were jostling around her, Bessie Bunter nosing in greedily, Gentle standing politely waiting on the fringe whilst anxiously glancing about her for her lamb, all of them nosing the grass in search of more nuts. After grub came the count, and the re-count, and the second re-count. Morning by morning a closer relationship seemed to enclose woman and sheep. Femaleness bonded them. She was in it with them, feeling in her loins too the burden of their pregnancy, observing them more keenly and more sympathetically than a mere man ever could. The approach of the next birth was always more certain to her now. That several were visibly heavy with lamb, that their udders were filling, this was plain enough to me, but I only observed it, I did not feel it. But she did.

The flock had never stopped going into the wood. We called in neighbours, and Haydn, Neil, Simon and I spent a laborious weekend 'laying the hedge', as we tended to boast of our rather untidy work. At the hedge's most fragile section, ash, hawthorn, blackthorn and hazel found themselves mutilated and tangled into the roughest of lines after our threshing about with chopper, axe and bowsaw, and into this, cosmetically, we wove an enormous amount of cut gorse. The ewes, coming fresh to it, acknowledged its existence and even its temporary efficiency by nosing along it in a quick inspection. But then, as we retired, leg-weary yet satisfied, they simply waited awhile, climbed higher up the slope, pushed and shouldered against other rotting remains of derelict hedging, went through, poised themselves on top of the bank and skipped nimbly over the rushing water, once more to graze and explore the quietness of Morgan's Wood. The difference now was that we no longer had to scramble round to head them off, but merely to rustle our bag of nuts and to call them back in our pseudo-Welsh.

"Tot, tot, tot . . . Come on. Tot, tot, tot . . . "

Very soon they would emerge, like refugees, trailing skeins of wool, with yard-long brambles entangled in their poverty-stricken fleeces, some limping and carrying their burden of precious unborn lamb with such swinging and careless disdain as they came that they seemed to have no idea of what was going on. Baa-ing, rewarded with nuts, they would stand around, once we had fooled them into a return to Dolgwili, hopeful of more nuts while they watched us set about mending that un-mendable hedge, both them and us certain that it was a waste of time and that tomorrow they would be through it again.

Gentle's lamb flourished. It took some time and not a little finger-counting for us to realise that the congratulations we had spontaneously showered on Sam-Ram for siring Gentle's lamb were entirely unearned. An ewe carries her lamb for about one hundred and forty-seven days; Gentle had conceived back at Llynwalter. Not until 21st March was our old gentleman's fertility put entirely beyond doubt. That morning, twin ram lambs were born to Spotty in the same sheltered corner high up on the Steep, again all easy and troublefree, for which we were so grateful and relieved that we gladly climbed that one-in-one slope three and four times a day to ascertain that all was well.

The next day all was far from well; not a sheep was left on the farm. Worry habitually creased my mind. Ever since that first day, I had worried that they would all drop dead, catch dreadful diseases, would take off and head back into the mountains, that foxes would tear them to shreds, that they would mix with our neighbours' flocks, that Sam-Ram was far too ramshackle ever to be fertile, that we had not enough grass, that rustlers would come by night, that our own townie dogs would savage the ewes, that this, that that, and so on for ever more.

That day Phyl worried with me. Shaken and let-down by her wonderful ewes, rustle her feedbag and call and search however and wherever we may, she stood with hurt eyes and admitted them vanished from the face of the earth.

Simon had been right, once more, of course. Even I had been almost right. Sheep on Dolgwili had always been a terrible mistake. Crazy, I had been, ever to listen to my Temptress. Mad, I had been, ever to have been seduced by those siren voices that used to sing to me of returning to my peasant roots, of living the simple pastoral life. Sheep consumed every minute of every day, sheep wore our legs down to the kneecaps, they·

42

worried us stiff and devoured our lives. No; Dolgwili should have been left to the rabbits and I should have kept to my gardening. Let them go. Let us cut our losses and forget them.

But Phyl was Charlie Kirby's daughter all right, and Charlie had never given up. Nor did she. She tramped and tramped and searched and searched. No matter how often I introduced the 'leave-them-alone-and-they'll-come-home' argument, no matter how often I suggested coffee or the police or phoning around or that Simon had been right, no matter how much I tried to push on with my gardening, persistently and obstinately she carried on her search – and so, of course, did I.

One learns. One learns by the exact degree of shininess on those neat black pills of dung just how long it is since they were dropped, and one learns to stamp a heavy boot on them to cancel them ready for tomorrow's search. One learns to carry secateurs similarly to cancel wool gathered on thorns. To save the hill-tortured legs, one's mind stores scraps of evidence of behaviour, memories of landmarks. The imprint of a hoofmark in mud, the memory of last night's weather and whereabouts of the flock at dusk, which neighbours' field holds sheep, the direction of the wind, all such things instruct the shepherding mind.

Yet, also, one does not learn. The cool logic which instructs is so often nullified by anxiety, by habit, by prejudice and by lack of imagination. Cool logic instructed us that we needed to fence all around the farm and that we needed a Juno, or perhaps even that the whole project was quite futile, but really there never had been much logic either about coming to Wales or sheep-farming. It was all instinctive, and having lived so long by instinct, we would have found it difficult, to say the least, to have been coldly sensible. Fencing all the way round the farm we could not afford, anyway, and the sad old boxer and the unreliable saluki on our hearth were unlikely to welcome, after all these years, a newcoming sheepdog. And to abandon sheep-farming now was unthinkable.

So, that day, totally entangled in it all, we walked and worried, examined the hedgerows and earth for evidence, pondered and cursed, until at last we were so tired we could do no more than ask Simon's help and he, in turn, asked Haydn's help, and they, late in the afternoon, found the whole flock at the further end of Mr. Thomas' farm up above us and quite a mile from our boundary. As Simon, having with Haydn driven them back on to Dolgwili, handed the flock over to us, he allowed himself a generation-gap look which said wordlessly to his doddering old parents: There, now just look after your things properly. I told you this would happen.

At roll-call next morning, however, Dainty was missing. Once again we started our tramping and searching, only to receive a belated and casual message from Simon that, Oh yes, there had been one close to lambing, so they had left it up there.

So, far into Mr. Thomas' farm we went, worrying once more, and there in the hedge sure enough was Dainty, with twin lambs, strong, safe and sound. Mr. Thomas, smiling his charming slow smile, waved away our apologies and could not have been more helpful. He not having the English, you see, and us not having the Welsh, you see, the conversation jumped about from misunderstanding to doubt and back before once again he waved away our thanks and we started our homeward procession, Phyl leading with one lamb and nuts, Dainty following her, me protecting the rear and carrying the other lamb.

Next day, two more ewes had twins. Lambing, we swanked to anyone who would listen, lambing was in full swing. It filled our thoughts and our dreams and our words. It was a fever.

My shepherdess arose first each morning now and was gone straight to the hill with her stick, her bag of nuts and her strange sense of fulfilment. Sometimes over my solitary breakfast some Freudian suspicion would edge into my mind for consideration that maybe she longed for another child. Day after day she walked and watched and counted and re-counted and fed and reported whilst I merely observed her happiness and her beauty blossom and acknowledged her as Boss.

"Another pair of twins, both ewes, born dinner-time today," the diary records, "the sixth pair so far that Sam has produced. We are wondering if we should not keep him after all."

Poor Sam-Ram. He had been an affectionate sort of joke ever since he arrived. On the day in winter when old John Jones had finally presented his account – a lot of funny little figures pencilled on the back of an ancient envelope, all of which added up to a colossal undercharge by him – we had stood leaning on the gate afterwards, as usual watching the sheep and ponies.

44

"Everytime he tupped an ewe," I reported hilariously to John Jones, "poor old Sam used to turn and look over to me and call out, 'Cor, my feet are killing me.'"

And old John had laughed that merry red-faced laugh and answered, "Yes, yes. Change him you'll have to, you see, next season. Yes, yes."

A stop-gap, Sam-Ram had always been considered. Yes, yes. Get a good young one next year, it had always been. And even Sam himself had encouraged the notion within us. At times, his feet were quite broken down, quite sad to see. Despite Simon's treatment of the condition, it became almost normal for Sam-Ram to hobble. Yet, for all that the books kept repeating to us that the ram is half the flock and a ram is only as good as his feet, here was poor old foot-rotten Sam now daily proving to us his magnificent fecundity, morning after morning.

Down on the low ground at evening we would scan the hillside. Invariably one ewe would be leaving the flock, almost imperceptibly, inspecting and then choosing her lambing site out of the wind, under a hedge or in the lee of the wood. Simon, after rabbiting up there with Urchin, would come in to supper with the confident assertion that Black Collar, or whoever, was 'just starting', or, playing on our old-fogey nerves, warn us that we would need to be out early next morning for he had seen a fox run through the flock just now.

It was all new and wonderful to us. Someone-up-there had passed the word that there were rookies at Dolgwili, that fair weather was needed, and easy births, and twins. Morning after morning, that was how it came. Usually the lambs were already there, lusty yet diminutive; sometimes they came later for, we supposed as we enjoyed these continuing smiles of Good Fortune, our particular benefit.

Then, within her chosen lambing area, the ewe would become increasingly restless. Token grazing alternated with token nest-making. She scratched tentatively at dried grasses, turned round and round as a settling dog does, did a little symbolic gathering of leaves with an elegant movement of her hoof. This nest she occupied soon in a characteristic side-lying pose, lifting her nose skyward and backward as straining started. Came a black head. Rest, more straining, until soon the whole package plunked out in its polythene-seeming bag and the first move-

45

ment of life in a cruel world. Nothing afterwards would ever be quite so wonderful again, except, in ten or twenty minutes' time, the twin. Cleansing by continual licking of the firstborn, a seemingly deliberate frustration by the mother of the lamb's first groping efforts to suck, the shaking, staggering lamb, the casualness of the ewe, all these delicious things we watched in a state of continuous euphoria, thinking how clever little we were, how virile Sam-Ram had proved himself to be, how splendid all our haphazard ewes, and how good God was – in approximately that order.

Nellie Eighteen, of course, had not yet lambed. The 'crazy, bare ewe', the diary records her as for immortality. She was still haring about the hillside, her fleece now almost completely gone, with only a shabby remnant adhering to her lean flanks, her neck quite bare, her eyes always apprehensive and her bleating as terror-stricken as ever. She really was a poor lost soul of a sheep, always in a muddle. We watched her with apprehension.

The dusk came, however, when Simon observed her hurrying off alone along the path through a large patch of bracken towards our wood. He had inherited from good old Charlie Kirby, his dairyman grandfather, a way of announcing with great significance that So-and-so would calve tomorrow morning, and although with Nellie Eighteen her dusk stroll in bracken might well mean only that she was lost once more, nevertheless her absence at next morning's roll-call almost certainly meant she was lambing.

Our wood was, in parts, terra incognita. That area closest to Morgan's Wood was fairly open and friendly, mostly hazel copse, but the part nearest the Steep, towards which Nellie had been hurrying last night, was solid blackthorn scrub adjoining solid gorse scrub. Phyl insisted – and I, still at breakfast, insisted too – that she would go up there.

She was gone hours. Times without number we had told each other that we simply must have some means of two-way communication for such situations, and times without number such means had never been found. What on earth was she doing, and where? When at last the bleak future which faced me, of always cooking all my own meals, became unendurable, I set off. That day, it turned out, everybody was looking for everybody else.

46

That patch of black wood was an original Disney setting for a coven of his witches, but the woman I discovered crawling out backwards had fair hair, however awry and decorated with black thorn twigs, had rosy cheeks, however scratched and muddy, wore anorak and trews, however torn and decorated with yellow mud, and the lamb she carried finally proved that this was no witch but my own tired yet triumphant shepherdess.

She had, some long time ago, discovered the lamb dumped in a mud puddle and possibly abandoned by Nellie Eighteen. Backwards and forwards and up and down, ewe and woman had moved about that awful wood in a crazy search and effort to unite lamb with ewe and to lead them out to the safety and comfort of the pasture. Even now, as I arrived, Nellie Eighteen had turned and gone back into the wood towards the place of birth, quite oblivious of the lamb in Phyl's arms. So we had to do it all over again, me crawling into the tunnel of blackthorn and on hands and knees getting round behind Nellie so as to drive her with shouts and the knock of my stick on wood, Phyl to cajole and call with the lamb in front, trying to thwart that strange connection of animal with place, of Nellie with her nest, which so intrigued me. For all her foolish choice of birthplace – she could scarcely have chosen a sillier place, or a more danger-ous one – as soon as we had given back her lamb on the grass down from the wood, Nellie Eighteen settled and almost seemed to come to her senses. She became immediately an excellent mother and thereafter devotion cured her craziness.

Day by day, my wife became godmother to lambs. The bonds forged as she anxiously observed their first minutes and hours tied her to them onwards through the whole season. She sat among them to watch lambs nuzzling and banging on swinging udders, sighed at the thinness of the ewes, delighted in the dancing capers of the growing lambs, noted the good and the bad mothers, counted and checked and worried into the even-ing and into the night. There was so much we did not know.

Crows were the worst danger, people said. Crows peck their eyes out. Crows, ravens, buzzards – and foxes. Foxes had been known to attack and rip open ewes in labour, people said. And badgers took lambs too, they said. And magpies. Phyl slept restlessly, with crows like Hitchcock's gliding down from that black wood with wicked beaks to attack her blameless lily-lambs. She ate small meals absently. Her make-up and her clothes, once so important, now were nothing. She was still tired at waking, yet each day was a new exhilaration. She was taken with lambs: she was in thrall to sheep.

How innocent we were. How little we knew, and how merciful the Good Shepherd was to us, his little lost sheep, that year.

Black Collar, one would have thought, should have known better, yet only two days after Nellie Eighteen's come-uppance, Black Collar led us a similar dance.

It was a beautiful morning. Almost from the very first day of the lambing, the spring had come good. Inspecting, noting this, examining that, counting and re-counting, Phyl did her early morning rounds before her snatched breakfast yoghourt and this morning the report was that Black Collar was missing.

"Oh Lord, not that wood again today, surely?" I groaned over a hot stove.

She thought not. Black Collar was sensible enough. No, she wanted no help, thanks. Black Collar was most likely just tucked behind a bush up the Steep.

Well, it was her show. There was much digging to get on with, splendid catalogues to leaf through, a great million gorgeous subjects to be planned into my border. She could always yell for help.

No yell came, but when she herself came she was leg-weary and mystified. Black Collar, now, had completely disappeared. It was the story of our life at Dolgwili. She was not up the Steep, not over Mr. Thomas' farm, not along the railway, nowhere in the hedges or in the woods. Nowhere.

The track that leads through Dolgwili from the road up to Mr. Thomas Penycraig-something-or-other climbs sensibly upwards at a most modest incline. Its foundation-stones must have been put down by countrymen a century before; mostly they had disappeared under grass as a century of rain washed down the soil from the higher slopes. The earth displaced to make the track had been piled to become a bank at the side, with slate stone walls set on edge to retain it, and hawthorn planted atop of that. All had been neglected over the years. The unkempt hawthorn met overhead and in winter blackbird and thrush threw themselves from the berried branches in panic at our approach. It was pleasant now however to walk up the

48

sheltered track on a morning like this and to lean on the red-lead painted boundary gate at the top, a natural resting place from which to gaze out across the fields for ever more dotted with sheep and Friesians. Ahead the track continued on towards Thomas' farmyard, hugging the hillside, whilst all down to the left the world fell away abruptly, over gorse, rock and bracken, to the shiny railway lines and the white cascade of river far below.

I stopped and stared, as everybody always stops and stares here. It is the top of the world up here. Raven and buzzard sit up here. Hawthorns yield their permanent shape to the west wind and the air seldom gets soggy with mist, as our Dolgwili air does when the mists rise off the Gwili to make the valley look just like a Chinese painting. Exactly why my feet should have wandered up this way I do not know, but I climbed the gate and went a little way further on Thomas' land, looked down towards the railway line and there, twenty yards down in the scrub herbage, was a sheep. Black Collar.

I called her Fool, softly, and returned over the gate, down a little and then through the hedge out on to the edge of the world. The sheep was about ten yards in from me, turning and turning again among the brownness of last year's bracken, nest-making. The slope was excessively steep, but this she overcame by scraping debris to the bottom of her nest. She was in no danger. She had not yet lambed, and we settled down to wait, both of us.

The sky was blue, the sun shone. A buzzard drifted over high above and on a white stone outcrop among the gorse an ancient raven sat watching nothing happen with little interest. Far below the river glinted among the dark trees. Soon Black Collar started the familiar straining motions and within minutes she had given birth to another fine lamb, and shortly after that a twin, allowing me, her locum shepherd, to sit in attendance and watch every splendid detail of the blessed process. The birth-place, which had seemed so unwise, turned out to be ideal, warm and sunny and free of all worry. I stayed until she had licked them clean, cut and cleared a better pathway for her to bring her lambs out, and came home. When Phyl checked later on, the whole family had returned to the pasture and both lambs were sucking.

That left only Scratcher. She was Scratcher because she had

49

a persistent habit of scratching her low-slung belly with her hind hoof, for no reason that we could ever discover.

When Phyl found her before breakfast one morning, Scratcher was already in a severely distressed state. She too had chosen an inhospitable site in an untidy hedge on the steepest and roughest part of the Steep. One leg was caught in between the branching trunks of a hawthorn bush, she was on her back and she was in labour with the head of the lamb just showing. Somehow, wrestling beyond her weight and over many rounds, Phyl managed to free the ewe but by that time both were pretty well exhausted. Searching the hillside through the binoculars when eventually anxiety touched me, I saw both ewe and wife resting on the grass about ten yards out from the hedge. I shouted questioningly up and she shouted answeringly back, but the words were lost in distance. They could only mean trouble anyway.

It was bound to come, of course, our moment of truth. We had been getting too cocky; we had to have our faces slapped.

By the time I climbed up there, the ewe was resting on her side, her body heaving with the exertions of the last two or three hours, her breathing very laboured. We could just see the black wet face of the lamb which was no nearer being born than when Phyl had first found the ewe stuck in the hawthorn.

We dithered. It is one thing to read the books, quite another actually to practise their advice. Hot soapy water, I remembered. Be very, very careful. Or fetch the vet? Or phone Simon at work? Either way would take an hour. The minutes hammered away at us. The ewe was far gone, straining occasionally only in a feeble and automatic way. We dithered a bit more, knowing that the priority had changed now. The lamb would be dead; the ewe it was we must save.

Phyl stood up suddenly and yelled, "Peter!"

Son-in-law Peter was just off to work, far below.

"Can you phone Mr. Thomas and ask him to help? We've got an ewe in trouble. It's urgent. If you can't reach him, phone the vet."

"OK?"

We waited. In no more than ten minutes or so the good Thomas was hurrying down to us along the slope.

"Duw, duw, duw," he said softly, shaking his head and

taking off his jacket. "Get a bucket of warm water and soap, can you?" he commanded, gently and assured.

I ran all the way down. I would gladly have run all the way back too, except for spilling the water.

"Duw, duw," he was still saying as, intent on the ewe, he washed and lathered his thick fingers, hands and forearms with soap. We held the ewe quietly on to the grassy slope. Thomas knelt and with great tenderness entered the ewe and brought out the lamb. It was dead. He threw it down. We held on to the ewe. He entered a second time. The second lamb was dead, too.

"Aye. Several days maybe, you see. Aye."

He towelled his arms dry. We thanked him and apologised for taking him away from his work. He lived alone up there now; his mother was in hospital and life must be hard.

His face broke into that charming smile, he shook his head disarmingly and softly said, "It's all right . . . "

He walked slowly back to his farm with his dog. Scratcher stood up and seemed to look a little plaintively after them. We sat silent. The ewe began to recover strength and to search for her lambs, baa-ing. They rested where Thomas had dropped them, beyond hearing. Soon Phyl collected the bucket, the towel, the grassy soap, and I picked up the cold wet lambs.

I think I do not know a sadder sight than two dead lambs. Lately I had planted a small new orchard. Now I buried one lamb to give eventual benefit to the Cox Orange Pippin, and one under the Blenheim Orange.

For two whole days Scratcher toured the farm calling plaintively for her lost lambs.

8   It was the mark of sheep

A stricken man regularly shuffled his slow way along the pathway on to the bridge high above the little river, there to sit
resting on the warm stone of the mightily-built parapet. This
was Griffiths. He pointed with his stick and his good arm up to
the Steep above our cottage. Oats, he could remember, up
there. Aye. He moved the stick along, pointed to the right of our
home to where bracken and gorse were continually gaining
ground in their invasion of the grass of the pasture we had
named the Rough.

"Potatoes there," Mr. Griffiths' reedy small voice informed
me. His stick made a horizontal stroke through the air. "A horse
plough, that way, along the slope, you see. Aye, aye. Potatoes."

To a lowlander, it was unimaginable. Merely to walk those
slopes tired the thighs, but to take a horse and a plough, to cart
seed, to cultivate and harvest a heavy crop up there, that was
almost unimaginable.

Aye. And the trains, you see. Busy, they were, all the time
then. Aye. Twelve waggons and teams of horses they had, him
and his father in the old days, working from the station here.
The hay and the straw and the coal, working all through the
valley, up to the farms, you see, him and his father. Busy it was,
you see.

"There's lovely," I felt like adding, picturing all the bustle
and the joy then. His eyes watered, and complaint at his use-
lessness now was never far off. All his life he had been so active,
but now . . . Aye, aye . . . At the confluence just along there
now, the best salmon that ever was taken in these parts Griffiths
had. Eighteen pounds, on worm. And the singing, you see. But
now . . .

Some days I would ask him if he had seen our sheep over into

Morgan's Wood again, but he never had, and soon, with conversation always uneasy between us – unfortunately, for always I wanted him to talk more of those old days of hardness and joy, surely – he soon would lurch unhappily away home.

Joy? No, perhaps not joy, but surely satisfaction? The picture that Griffiths' few words conjured contained so many more people all rubbing shoulders hereabouts to make the valley work then, the regular steam trains pushing alongside the river through the hills from sea to sea, the farmers, the merchants, the children, everything so local, everything so interdependent, so known. Perhaps nostalgia over-colours the picture, but when I compared what was then with what was now, it was difficult not to sigh and wish that one had lived here then. Now the motorcar and the English were everywhere; all was changed for ever.

For ever? That was a big word. Perhaps it would all go back to what it used to be. Man was stopping the machines. Man was staying away from the factories, the mines. Man was deciding that no money was enough. Man was walking out, turning back. Man was turning again, in however a fanciful way, to watching birds, to fishing, to keeping sheep, to ploughing with horses, to singing, to crafts. Perhaps Simon's grandchildren, too, would see again twelve waggons with teams of horses loading straw at our little station and he would sit with them on the bridge to tell them how much better it all was . . .

That bridge was a magnificent structure. A single span of dressed stone stood high above the tumbling river to carry the railway from Carmarthen Bay in the south to Cardigan Bay in the west. Now even the last old milk train had stopped running and at last we did not have to worry that our lost sheep would be decimated, as Tiger Lil had been, down on the track.

From the bridge, all of Dolgwili could be scanned except for the very top left-hand corner and the bottom left-hand corner, both just hidden by the lie of the land. From down here on lambing mornings, Phyl could always clearly be seen ascending the Steep by slow zigzag, like a climber on the Eiger, and from down here at evenings any ewe isolating itself from the flock ready for lambing could easily be spotted. From here, through binoculars, the grey shapes of our stray sheep moving among the black trees in Morgan's Wood could be picked up, and usually the whole flock could be counted from here – as long, that is, as one used binoculars. The face of the Steep was pimply with small sheep-shaped and sheep-sized anthills covered in early spring with grasses and mosses still winter-dun, which was the colour of our sheep now, so that, without binoculars to differentiate, one could clearly count about sixty-eight sheep.

The counting of sheep may well be an art. One count is never enough, four is merely confusing. And it does have to be right. Sometimes we count and count and at the end of it all one is never sure if poor Nellie Eighteen has gone walkabout once more, or, if, perhaps, Darkie is an anthill or an anthill is Darkie or what about that one just gone behind the gorse bush, so you count again. And again.

"I can't get to sleep, Mum."

"Of course you can. Count sheep jumping over a stile," she used to say. Ah, if you could only see me now, mother.

Perhaps we did not use the right words. *Ain, tain, tethera, methera, mimp, ayta, slayta, laura, dora, dik, ain-a-dik, tain-a-dik, tethera-dik, methera-dik, mit, ain-a-mit,* didn't that sound a lot better? Would they not get counted more accurately that way? Or, since they were Welsh sheep, *un, dau, tri, pedwar, pum, chwe, saith, wyth, naw, deg, unarddeg, deuddeg, triarddeg, pedwaraddeg, pymtheg, un a bymtheg*?

Merely to see sheep was to count. And, like that great flash of pleasure which greets the sudden speaking of a truth, so too a flash comes quite intuitively when one knows that one has counted correctly. A little tune issues from your mind and trickles down to your lips. You find yourself sitting on the ginger-mossed stone in the morning sunshine, humming a little: the sheep are all there, all is well on Dolgwili. You can stay here awhile, just listening to the river below, just wondering if that is a willow tit and when will coffee be called.

The white cottage squatted long and low, plumb in the eye of the sun, plumb in the centre of Dolgwili. In front of it, right down here to the river, stretched the undulating acre-plus of sweet grass which was the garden-to-be. For a focal point it already had a great fallen elm, one which had got its feet wet by being planted in a saucer of rock and been blown down last year in the early summer gales. Now its ancient black boughs were being rejuvenated by the bright green leaves of spring.

Right across the grass, inerasable from west to east, ran a scarring dark line. No matter how many times the mower crossed and recrossed the grass in its long lawn-making, the line remained. It was grass, but it was dark grass, darker than anywhere else.

It was the mark of sheep. It led from the gateway of the meadow we had named Sheepdip and was a direct line across to the Spout, drinking place for generations of sheep. It then led out by what had been another gateway to the Rough. The line was the mark of a million sheep, footprints of centuries perhaps, and to me, garden-making, its permanence there hinted at their ultimate return. The fine grass nurtured by their droppings perhaps had been merely lent to me for these few years, a plaything to be given back. Already their descendants, our own Specklefaces with their darling lambs, were threatening the garden, hinting along the hedgerows, gallivanting at the gates ready to come back in and tread that age-old line.

The sheep had the run of the farm, except for my garden acre. Sixteen ewes, even with their lambs and even on our uncared-for land, on twenty-something acres was far from excessive stocking. They should have been content. Yet always they were pushing out, searching for the greener grass on the other side. Every night they always made their way up the Steep to the shelf at the very top, thus receiving the last warmth of the westering sun, thus avoiding the valley frosts. At morning invariably they descended diagonally to the Rough with perhaps a drink at the Spout before crossing behind the cottage to their favourite grazing on Sheepdip. By late afternoon they edged up again towards the Steep once more, whilst if it was stormy they entered the wood. They had the run of the farm, but they were never satisfied. You could sit here on the bridge and through the binoculars you could lip-read.

"Sisters," Spotty the Disruptive would be saying between mouthfuls, "conditions are terrible here. Let's try Thomas'. He's got a bit of reseeding the other side of that old hawthorn hedge. Come on, sisters, let's have unity now," and off the lot would move, scrambling up the bank, heads down, pushing with determination, hellbent on truancy. I would leave my musings and poor Griffiths, dash home, grab my wife, my heavy leather gloves, my deerstalker, my bowsaw, my long-handled

55

secateurs, my knee-pads and a bag of nuts and head outwards for the break-out.

One day Griffiths would have to instruct me on how they managed in the old days. How did they keep their sheep in? By now, Dolgwili's hedges were poverty-stricken things, with thirty or forty years of neglect showing in every one. Gaps were filled with old bikes, old bedsteads, old milkchurns, anything. Rusty old wire, even old hawsers, snaked and roped among the hawthorn and hazel, some embedded deep into the flesh of the bushes. Dumps of old bottles, rusting tin cans, great old bones from the stewpot turned up in every hedgerow that was any-where near the home, and nowhere on the holding was there any worthy fencing except at the bottom along the railway and that, once marvellous, was now getting untrustworthy in places.

Hedge-mending, every time you set out to do it, is perfectly straightforward. You lug your wife and all your gear up there – it is almost always up – you regain your breath and your legs and then either you push her through or she pushes you through. You don't, after all, want to enlarge the hole and thus make more work for yourselves. So you shrink yourself and crawl through. Once through on to Thomas' land, act casual. Hum a little. Speak to the flock. Wander round a little and admire all of Thomas' other hedges which obviously have been trimmed and laid regularly over centuries. Only then should the bag of nuts be produced. Offer Spotty one or two and so get her hooked. Move slowly, ensuring that all ewes and lambs are gathered before you move off and crawl through your hole back on to home territory. Keep them coming and take them down a way before scattering just sufficient nuts on the grass to keep them engrossed whilst wife and you seal off the hole with anything to hand. Old bedsteads are very good, or old milk-churns, bicycles, anything you happen to be carrying. Now you can start your shouting and swearing and arm-waving, stone-throwing, cursing, anything to send the damned sheep away from their remembered break-through point. Only then can you relax slightly, sit down and look across to all the other immaculately hedged farms, and wonder how on earth it has come about that you are here with her up on a Welsh hill worrying about sheep.

56

"Are you sure you still want to keep sheep?"

And she answers, "I'll go and cut some gorse," whilst her Adorables gaze at us sullenly from a distance and plan tomorrow's Operation Break-out.

Invariably hedges are grown on banks. This automatically bestows the fundamental disadvantage of a sloping foothold. If it is wet – and it always is – one's feet slide away backwards continually and many is the time we wished our feet starboard-and port-pointing rather than bow. Ideally all the trash should be cleared, the blackberry or gorse or seedling hawthorn outer defences that sheep come to before the main rampart of hawthorn or blackthorn. So you chop this all down, suck the blood from your wrists, pull bramble tentacles clear of your trousers and try to define the break-out hole. Into this you now push terrible boughs of thorn cut from the top of the hedge. These, as they are cut with the long-handled secateurs, will certainly fall straight down on to your crazy face and your foolish head, but no matter, each piece is a prize and your hole is becoming quite impregnable.

Nearby, however, there is certain to be another part, weak now in comparison, and bit by bit you will be enticed all along the whole blasted hedge, mending and strengthening until even you have to admit how exhausted and thirsty and shaky you have become.

"That's enough for today," one of you had better quickly say, and so, satisfied with a hard job well and truly done, you can sit and rest awhile. As you sit you may well hear a small tearing sound. It is nothing. Merely the sheep further along, pushing and making a new way through, chuckling slightly as they go.

Why exactly did the old folk – those splendid old folk of Griffiths' day – why did they build a bank, reinforce it with stone and then build an untrimmable hedge on top of it? They were wide hedges, high hedges and how on earth they cut and shaped them in the days before tractor-hedgers puzzled and annoyed us.

"Look," somehow I found courage to say when I realised how much all this hedging had aged me, "we can't go on like this."

"That," she stood back from her latest effort and boasted admiringly, "that should keep them in."

"*I* can't, anyway. Just look at me."

"Had we better check it higher up?"

"If I'm going to die, I want to die in the comfort of my shrub border. Somewhere close to the blue atlantic cedar, not out here on this bleak Welsh hill."

"It's not even that the grass is greener on the other side, is it?"

"No, it's just sheer bloodymindedness, that's all. They know we're English. They're Welsh. Listen to them laughing at us. Beulah Speckleface! You can't expect to keep that lot in. What's the use of breeding a hardy sheep that will fend for itself on the wet and windy hills and then expect it not to break out in search of food? You and me is crazy, gel, working like this. Y'know, I don't really enjoy hedging at all. I don't want to spend my days up here mending dirty great holes in blasted hedges. I want to dabble in the sweet loam between my shrubs, with the perfume of lilium regale in my nostrils, the buzz of honey bees and the erstwhile melody of a friendly robin in my ears. Besides, it just can't be done. Better men than you, my love, couldn't make a job of that hedge. Hedges have to be laid. It's a skilled craft. Hedges are constructed, built thick, impenetrable, even beautiful. And then serviced regularly generation by generation."

We glared at each other a bit. Soon she wandered away to snip off some more gorse from the scrubby growth infiltrating the pasture, gathered it in her poor arms and came and pushed the prickly stuff at another imagined weak place in the hedge. She was far gone in sheepness. What's more, she was Charlie Kirby's daughter.

"If," I pronounced, "we're to keep the damned things, let's keep them properly. Let's get the place fenced and hedged before next winter, so we can rest easy. As it is now, behold the winter is past, my love, the singing of birds is heard in the land. Very soon now the new grass will be growing and perhaps they'll stay home for the rest of the summer."

The spring had been cold, the growth of grass mingy. On all the farm, whether viewed from up here on the Steep or right down there on the bridge, only one spread of grass showed greener and lusher than anywhere else: my lawn. Twice a week without fail, I trudged backwards and forwards behind the rotary mower, throwing off a green confetti of grass tips back for

58

the lawn to feed on, hour after hour, creating that fine sward into which would drop those long-planned islands of black earth swarming with peony, iris, delphinium, conifers and rich brocades of many-coloured foliage. Daily now around the gate of Sheepdip, Spotty and her Disruptive Elements gathered, bleating to be let in to their birthright, my lawn. The blasted darling lambs were already crawling through between the gnarled hawthorn trunks and trash of the Sheepdip hedge to nibble my lovely grass, whilst old Sam-Ram stood like a union leader, vociferous yet still dignified, with that deep and plaintive baa in his demands for decent grazing conditions for his wives and kids.

For lambing was now over. Sixteen ewes had produced, with a little help from Sam, twenty-four lambs, all thriving. That, a lambing percentage of one hundred and fifty, was, considering our ignorance, almost miraculous. Only the stillborn twins grieved us still; we should have risen earlier, we should have been more attentive, more decisive. Next spring, now . . . Ah, next spring.

Meantime the ewes were hungry, and my wife was inclined to line up with them in support of their daily demonstration outside the gate. Certainly it did begin to seem foolish during my three-hourly stints behind the mower to be killing myself at the cutting and wasting of grass for which they stood outside absolutely drooling. This, however, was my garden; this was why I was here. The garden had always been in the scheme of things, the sheep never had. This whole acre was my waiting canvas, soon to be filled with gorgeous colour and immaculate patterns. Already seven neat squares had been cut into the turf, filled with peat and compost and now stood planted with seven dwarf rhododendrons. And down there . . . And over there . . . And here . . . Ah, what plans there were in my head then!

I held out against the inevitable for another week or two. The lambs daily insisted on pushing into the garden, and although the holes they made could always be filled with old watering cans or odd bits of chicken wire, I knew I was losing. The ewes, in any case, whenever they heard me insisting on the sanctity of my garden, simply turned tail as a flock and straightaway walked clean off the farm. You could see them think vengefully as they went, Right then, we'll show him. And that would be the end of my gardening for the day as once again I downed garden tools, upped hedging tools and set off on one more shepherding and hedging safari.

Finally I gave in. We erected a temporary defence around the dwarf rhododendrons and all along what other plantings had so far been achieved, opened the gate and watched them storm in. The ewes behaved superbly, showing no interest in anything but my lawn, over which they spread themselves with gusto. Sam-Ram, however, had interest in nothing else but my rhododendrons, and all the lambs had no interest in anything but my pinks. Why a ram of his breeding should ever be hooked on rhododendrons I do not know, but he simply could not resist the hard stuff. He stuck his great black head down and charged pathetically on his broken-down old feet like some sheep-Quixote at the conglomerate defences of chain-link and black-thorn boughs, until once again he could sink his poor old worn teeth into that forbidden fruit. My gardening became a daily routine of chasing lambs off burgeoning pinks and Sam off my poor rhododendrons. He would stand there afterwards with a reproachful look in his eye and pieces of black twig hanging from his tatty fleece, waiting for me to go in to tea so that he could start another charge.

May, however, despite all our earlier doubts, finally came. The pastures grew lush and satisfying, ewe and lamb rested to cud in fair weather. My garden acre was returned to me, the awful yellow wire of the old electric fence together with its rusty and inelegant posts were stowed away and we could sit back in respite at last.

What on earth were we at? Yet how on earth could we stop? We were caught up in a flow of nature. It was like being on a vast flood of some mighty river; there was a grand inevitability about everything. All one could do was to keep oneself and one's family safe, observe with interest, and go on.

9   Silly old fool, stupid young twit

The chestnut mare was upset that morning, Phyl reported.

"She has probably started then," I blinked at her sagely. Why do we all say such things with such knowingness, such grandeur? It was Charlie Kirby and his calves, Simon and his lambs all over again. As if all of us in turn had achieved importance and prestige by the vast significance of the impending birth.

Fleur continued to be upset for a long time. We stood off, watching her in turn. Some idea had taken hold of our mind that we must not worry her by being near, so we took station on the high ground of Sheepdip whilst the pony mare occupied a strip on the low ground down by the railway fence. She whinnied regularly, paced up and down, grazed, looked about her with ears pricked, grazed, whinnied, went and had a look over the fence, grazed a bit more, and so on, from pre-breakfast into the early morning. The assumption was that she was getting ready to foal. We waited and watched.

Unquestionably she was unsettled. Unquestionably we were puzzled. We were not pony-people, although when we did our sums correctly we could certainly claim to be more pony-people than sheep-people, for we had kept ponies before. Nevertheless we were always very aware of our inexperience, and very anxious over Fleur for she had cost us two hundred pounds and had been covered by a stallion of some repute locally.

Through the binoculars I discovered something on the grass, previously unseen by us, since it was hidden by a fold in the ground contours. It was almost certainly afterbirth. Where then was the foal? Perhaps she had had it in some other part of the farm. Perhaps it had not been born yet. Perhaps this, perhaps that. We walked the farm and found nothing. The

61

other pony, Syndod, was with the donkey and neither of them was particularly concerned, as by this time we certainly were. Keep an eye on her and report back, the vet said when we phoned. We could do no other than keep an eye on Fleur. Her behaviour was still as before – unsettled, worried even. Grazing, then a walk, the lifting of her lovely head alertly to look about her, a whinny, back to grazing.

Nothing so far in the way of routine work had been done; I decided to walk the dogs. It was June, but hailstorms kept blowing up in a cold wind and I left Phyl huddled among the gorse still watching for the foal to be born.

The milk train had trundled away back into history, but the railway track remained and along it ran a footpath, created by a free-and-easy community. Fishermen used it all summer, families blackberried here in autumn and a few of us walked our dogs and watched for birds along the well-wooded track. Sam-Boxer, Caramelle-Saluki and I had walked a quarter-mile, when a pale ghost of some strange animal – it might have been a unicorn – rose unsteadily from a damp place on the side of the railway track, among trees. I stopped. It was the foal.

It staggered and half-fell back into the black, leafy mud and from it I looked up the cliff face of hard rock, sheerly perpendicular, up to the level all of forty feet above, where our Sheep-dip pasture and the railway fence was. Somewhere over and up there, Fleur was still looking for her foal, as, indeed, she had probably been doing for hours now.

I turned and trotted home with the dogs, shouted for help and trotted back to the foal. She – yes, I noted, it was a filly – could hardly stand. Her thin legs buckled constantly and she had a tear in her golden skin above the knee. Phyl and Trudi cosseted her into a blanket and we carried her home. The hailstorms continued and Fleur was still in the same place looking for her foal, which now we presented to her out in the meadow. Soon we persuaded both back into the comfort of the stable and it all ended happily enough, but when I examined how the foal could possibly have fallen right out of the meadow and down the cliff on to the railway track I was brought up against all our fundamental problems anew.

Mushroom, as the foal came to be called, had evidently been born in that sheltered corner of Sheepdip and as Fleur cleansed her she must have pushed her along the slight slope down towards the fence. Hereabouts the fence was good. Great fifty-year old posts, six inch square and five feet tall, were as solid as when they were put there. The wire had been replaced fairly recently and to look at the fence was to conclude that it held no risk for stock or possibility of their passing through. Yet the foal had gone through a space that could not have been more than nine inches, and exactly at a spot where no bush or tree grew the other side of the fence. Mushroom would have plummeted directly down a forty-foot drop, must have at least grazed and bumped herself a bit on the way down, but fortunately landed on the soft ground where drainage water was held up by the raised railway track.

So once again mealtime discussion tied itself to the need for good fencing all round the farm and the utter impossibility of supplying the need. We could not afford it, we could not erect it, and even if we won the pools tomorrow, well, the animals would still get out. No; all we could possibly hope for was to limit them, to get just one side made impenetrable – Morgan's Wood side.

"Somehow we must do that side. I've spent positively my last winter chasing sheep through Morgan's Wood."

"Caleb and I will do it," Simon offered with his sudden smile.

"*You* will?" But then I saw the way the wind blew. "Ah, well, how much?"

"Forty quid," he said off the top of his head.

"Done," I said off the top of mine.

"You provide the wire," he provisoed.

"Of course. But you provide the stakes."

We glared at each other as father and son should once they have achieved, without real rancour, the 'silly-old-fool, stupid-young-twit' stage fairly painlessly. Both of us recognised where we were at, knew our place, him his side of the generation gap and me at mine. Mine, of course, was the preferable and right side, especially at a mere forty pounds. Young men have no judgment.

"But it has to be guaranteed sheep-proof, mind. I want a neat and tidy fence, with truly perpendicular stakes and straight wire."

With all the magnanimity of one who has struck a good bargain, and furthermore realising that this was the fishing season and that his mind would be filled with leaping fish until October, I added, "I don't mind when you do it so long as that whole side of the farm is sheep-proof before winter."

We inspected each other further.

"What about the foot-rot?" he asked, smiling that damned beautiful smile because of his strength and his power over me.

"What about it?" I asked with all the innate suspicion of a Suffolk man and knowing, too, that I was not as yet willing to acknowledge his power over me. They have no judgment, young men, but they do have the strong backs necessary for the treatment of continuous foot-rot.

"Ought to be done properly," he proclaimed. He was by now a fully trained farm-labourer.

"It certainly should," I sparred. The ewes were still walking-wounded.

He considered for a moment and confessed then that he and Caleb had been talking it over. "If you like we'll look after the sheep all the time for you, if you and Mum would let us keep a few sheep of our own."

Well, well . . .

"Would you now?" I exclaimed in delighted surprise, my new garden suddenly blooming in my mind's eye. "Nothing would suit me better. But I thought you disapproved of our sheep?"

Simon shrugged. He had the gift of silence. It was priceless. It persuaded people that he was deep, that he also had the gift of wisdom. But it did not persuade me.

I walked again Dolgwili's boundary from bottom to top where it ran with Morgan's Wood. As I panted and rested my way up, estimating the amount of wire, the certainty grew that at forty pounds, the fencing project was a best buy. I imagined lugging – or even cutting first – all the stakes. I imagined all the back and forth journeying with sledge-hammer, rolls of wire-fencing and tools, all the preliminary clearing of trash, trying to unroll the infernal wire among old hazel stubs, the need to peg the bottom of the fence to all the varying contours of the bank, to make it tight and decent, to find a good line, even remembering that there always had been a line of some sort down here. At forty pounds I was guilty of a rather pleasant form of daylight robbery. Really you needed a specially developed calf-muscle to stand on the wretched slope at all. Not a single yard of level ground existed along the whole stretch; in wet weather it was nigh impossible to walk up there at all, let alone construct a fence.

There in the hazel copse I faced the reality that without Simon there almost certainly would be no fence, without the fence there would be no sheep. The only thing that might stop them gypsying off all over Wales would be their chronic foot-rot and how on earth anyway would we cope with foot-rot again, but for Simon?

Foot-rot lames, foot-rot spreads, foot-rot stinks. To treat it, the textbooks say, turn the sheep on its rump, pare the infected parts of each hoof down drastically with a sharp knife so as to expose the rot for treatment chemically. To turn the sheep, simply roll it over one knee, pushing the head and the flank. Hold the sheep still, cut deftly, squirt or paint accurately, and the job is done.

Not our job. Our ewes had almost certainly never read the book. There is still no recorded instance of a single ewe of our flock ever having been rolled over a knee. Not only had the ewes not read the book, neither, it turned out, had our new Sunday shepherds. Their technique was soon revealed to us to be a sort of dance, of taking their ewe partner by the scruff of the neck or by the fleece haphazardly, or sometimes the front shoulders, lifting the ewe or throwing it off its back legs whilst at the same time pulling it backwards, hugging it and kneeing it forwards below. Occasionally an absent-minded or very naive sheep might find itself thrown fairly helplessly on its back or side on the concrete floor and quickly straddled and threatened by a sweating young man with a sharp but small penknife, hoof by hoof, but by and large there were few absent-minded or naive tenderfoots among our good old Specklefaces. Perhaps they had read a different book, for they themselves had developed a counter technique. At a precise moment they could twist and caper on their two back hooves, so that sheep and man tended to totter around the shelter in a rather desperate and ill-balanced embrace, falling about wild-eyed before a pause for rest and another round. Speckles are not large sheep, nor particularly heavy, but certainly somewhat agile. Simon and Caleb proved them to be most adroit dancing partners; the performance went on for several hours on a whole succession of Sunday mornings as the foot-rot was still not overcome. Their assistants, Phyl and I, gleefully watched these inelegant caperings from the doorway where we lurked with penknives and aerosols at the ready for when our sweating young shepherds should stretch forth a hand towards us.

It was always an unpleasant business. Simon and Caleb, however young and strong their backs, always suffered aches and pains from leaning over their capsized sheep and struggling to hold and treat each foot in turn, whilst the ewes always limped away from the shelter worse than when they had come in. The floor was littered with muck and urine from the incarceration, the smell of sheep adhered to everybody through the coming week, the problem of rot remained.

Worst of all always was Sam-Ram. His ramshackle – surely the word had been minted for him – shoes seemed unimprovable. At pasture he would stand holding his near fore off the ground or sometimes his back feet would be so unendurable he would graze from a lying position. He would have to go, we would decide again. Yet he had produced this magnificent crop of lambs. Perhaps if we had another good go at him next week? And so on.

Summer had brought a sheep-lull. Grass a-plenty kept ewe and lamb content. True, the ewes limped still, Sam-Ram rested more and more and occasionally a lamb hauled itself through to chew a rosebud or two but by and large contentment had arrived. Simon and Caleb had little enough shepherding to do and anyway the arrangement, such as it was, was due to start only when they had purchased and brought home their own sheep. Within the sheep-lull I could garden to my heart's content; only those Sunday morning foot-rot parades called me away from my digging, my planting, my mowing, my dreaming.

Too many of my precious hours passed in mere garden-dreaming, in pausing to ease backache by leaning on a fork to visualise with half-closed eyes what this miserable little patch of doubtful soil, with its scarcely discernible spattering of infant plants, could ultimately become. Perhaps those hours of seeing with the mind's eye all the glories to be would have been more profitably spent in garden-study, for whilst much was quickly apparent – that the soil was acid, heavy and slow to warm up, that our climate was mild and drainage good – nevertheless so much more was not apparent at all.

The seven dwarf rhododendrons from Bodnant, for example, which the mind's eye had assured me were beautifully sited in their peat beds on the brow near the gate into Sheepdip, and which had already been savaged by Sam-Ram, soon turned out to be disastrously placed. These were early-flowering shrubs, but frosts came fiercely all through springtime and sometimes well into June. So our innocent rhododendrons froze through the dawn hours and on until nine o'clock, at which time the great sun came surging up over the hill of Morgan's Wood to focus its hot beady eye directly on them, scorching them even unto death. One needs so many years to learn a garden, a house, a river, a person; it is hopelessly unfair.

What was apparent beyond all doubt was that the main layout of the garden had long ago been decided, as so many crucial matters are, by sheer circumstance. A whole series of builders, sub-contractors, sub-sub-contractors, moonlighters, men 'on the lump', occasional and anonymous labourers had all struggled here, some marvellously, some incompetently, some heroically, not only with our natural difficulties of access, terrain and weather, but with their own difficulties of temperament, finance and calling. These various gentlemen had bulldozed on, manoeuvred across, mixed and abandoned into, covered over and got stuck on, and generally so scarred the area close to the dwellings in all their brawling encounters with cement, sand, hillside, rain, stone and human frailty that whatever garden eventually arose from their mess must for ever bear their marks. They were strong, laughing men, almost all of them, but so tangled in the difficulties of building that the idea of a garden existing here probably never crossed their hard minds.

Beside the farmhouse and on the same hillside shelf had stood a modest barn which, from the outset of our family migration, had been earmarked as the future home of our daughter and her husband. Back in wartime, when Dolgwili would have been so unkindly directed somehow to grow its quota of potatoes and oats, this barn had no doubt housed the horse, a few cows and the standard pig and hens. Consequently it was certain that the vestiges of a tiny garden we had discovered nearby would have been enriched with manure surpluses, and that all the hummocky forest of nettles marked the graves of muckheaps of vast richness. These residues and the clean sheep-pasture in front were the only elements of virtue and promise I had at the outset from which to construct my dream-garden. Alas, they were to disappear beyond trace.

Both farmhouse and barn had for long been gently foundering into the soggy hillside, against which indeed they had deliberately been built a century and a half ago. Dampness was marked inside by a brown shoulder-high tideline along the whitewashed walls, and both buildings cried mutely yet desperately for air, air, air. A great yellow JCB brought them air by gigantically clawing out the slaty hillside, on which rested the vestigial garden, and then massively pushing tons and tons of this rocky sludge outwards onto the sheep pasture. A plateau of subsoil and rock took huge shape, yellow-grey, impervious and irrevocable. Worse followed.

The barn, of local stone, no doubt annually limewashed, and with the usual blue slate roof, was a thing of remarkable though modest beauty whenever the sunlight glanced obliquely off the cobbly irregularities of those white walls. That beauty in so commonplace a building was a celebration of the fine labours of those anonymous workmen who quarried, dressed and carted all those waggonloads from the quarry just down the road, across the moating river up to this man-made level in the middle of the long-sprawling hill. Who could bear to demolish such a building, for any reason whatsoever? The roof timbers were sound, the slated roof quite perfect, the stone walls intact.

Yet it was a desperate period. Two generations had been cooped up too long in one small room of our cottage, our lives were fraying away in delay, money was tight, circumstances ruled. And the one good builder we did find had to be persuaded, just as a preliminary difficulty, that he must ferry all his supplies across a flash-floodable river. In the face of all this, our fine ideas of architect-supervision to raise the roof or lower the floor, to use only native materials, to conserve and improve, had to be discarded. Bulldoze it, the new and callous generation of young builders cried. Start again. Breeze blocks, now, they're the boys.

So, insensitive to our sighs, they began singingly to demolish the barn, to pile an unknowable conglomeration of broken glass, old drainpipes, shattered concrete cow-standings, old doors, tiles, mortar rubble, window frames and, worst of all, masses of that splendid old dressed stone, all mercilessly knocked over and shoved to a rough level which extended even further the previous yellow-grey plateau. On it they shamelessly mixed their concrete. Mixer-rinsings, abandoned half-sacks of cement, piles of sand and ballast, sheets of polythene, cigarette-ends and crusts and swearwords all hardened into a thick skin yards wide and inches deep, a wasteland which ended in moraines and screes spilling down a twenty-foot slope to the original sheep-grass and covering in part that dark green pathway-mark of sheep. No matter what, the builders' plateau was there for ever more.

Powerless, and even at last grateful, we watched the old barn go and our daughter's home come with little thought then for any garden. We wanted only to be allowed to get on with our lives; get rid of the builders and let us have a little peace here, we cried in our prayers. Builders dwindled, builders departed. We stayed our grumbles, waved them farewell in relief and gratitude, set about completing or improving their work, and at last turned our eyes gardenwards.

Into that unholy conglomeration went my first unplanned plantings: amelanchier canadensis, cytisus praecox, a handful of unknown conifers grabbed indiscriminately at a garden centre, an unkillable peony which had survived the original garden massacre, anything, and that quickly, to disguise the daily awfulness of builders' debris.

The JCB, when it had crossed and recrossed the grass below the cottage, had by so doing decided the site of my main border. Where it had stuck and stuck again and clawed itself free, a long crescent of spewed turf and black earth soon insisted that it too was as ineradicable as the plateau and the screes.

Lessons lurk on a dusty shelf at the back of the mind. One knows they are there, but wisdom only arrives when one dusts them off and uses them. How can one not perceive that nature is master and mistress of all? Who, after all, are any of us to confront it, with all our cleverness and trickiness? How silly I had been with my pretentious neat squares of doomed rhododendrons, daring to think that I could decide the shape and scope of my own garden, as if I were the boss. At the back of my mind I had known all the time what I must do, yet I simply did not have the wisdom to sit quiet for a few days, to observe, to feel, to think into what nature was already doing here, to see what she would allow me to do. I had to go with her, to put my small hand in hers, to take it easy, to get to know her, to accept all the bounty she had to offer, aware of her reasons, her moods and her power.

Well, who does not know that? Ah, but to do it, day by day and hour by hour out there, leaning on the fork, to know just what to do, to be aware in detail exactly how to accompany her, there lies wisdom. And how long does it take? And how short is life . . .

So that was the way we tried to travel, the garden and I, in endless fog, peering, listening, searching for the way nature decreed we must go. If she decreed that slugs prefer tulips, very well, tulips must move from garden into tubs, and if rabbits preferred above all else, as mine did, pinks, phlox and broom seedlings, well, then, you threw up your enraged arms in almighty anger in the presence of your hunter-son and you blazed, "What sort of hunter are you, anyway? Day after day I come out here to see your cheeky rabbits running off with my seedling broom in their blasted teeth and all you can do is to give me explanations about rabbit-territory imperatives, that as soon as you shoot or snare one another moves in . . . "

I was very frustrated. "I tell you, mate, hateful as it is," I added, "I could just about stand a healthy dose of myxomatosis in this place right now."

He stared at the silly old fool as if I had blasphemed or as if I had slapped his face for the first time, or as if a damned garden was of any importance anyway. Five days later, God forgive me, I found a ghastly carcase of rabbit dead indisputably with myxomatosis on the track just at the back of the cottage. He could not possibly think that I had put it there, surely, but of course it spread like wildfire and wiped out most obscenely the entire rabbit population of the valley within weeks, but he looked at me sideways more than once that summer, that summer of growing pinks, phlox and broom.

Garden or farm, we all worked in uncertainty, in persistent fog. It was a bit like map-reading; all the requisite knowledge was set out on paper, yet reality was almost unrecognisable.

Simon, a full-time worker on a teaching farm after a year's practical agriculture in Dorset and a further year at farm college, plus Caleb, as near as dammit the Welsh nephew of a famous Welsh sheep-farmer in the Black Mountains, surely together they knew exactly what they were doing with our sheep? Yet there we all were, again and again, driving limping sheep down off the hill for yet another Sunday foot-rot ritual, with still that same old question mark hanging over Sam-Ram's noble head. Did they really know what they were doing?

Both were devotees of old Dai Sheep, it seemed. Simon certainly was, and unsurprisingly so. Dai Sheep was by no means the first corduroy idol in our son's life. Simon's ideal man, long observation had informed me, was any quaint and preferably ancient countryman with a store of poaching tales, or funny-gruesome stories, who could swear with natural grace, who wiped his nose on the back of his hand more than somewhat, who talked dialect and who had, somewhere at the back of his cottage, a shed of some sort hung with drying skins of stoat and rabbit, a few fox brushes, snares galore, hazel sticks, traps, old coats and boots and tools, a gun, an earth floor and who inhabited the place with accidental but genuine eccentricity. A den, it would be, totally male. The dog – a cat was allowable, too – would have no name. 'Dog' it would be called, or 'Cat'. Death was never far away from the den; dead chickens, dead fish, rabbits, hares, pheasants, foxes, crows, lambs, would all hang there while ancient and modern stood to chat haltingly of hedgerows, rivers, neighbours with spasmodic laughter. Dai Sheep did not conform totally to this ideal, but he was thereabouts.

Dai, Simon suddenly informed us, had some good sheep for sale. Good'uns, he earnestly assured us. Dai Sheep knew everything about sheep; he had lived his life in the flock, one might say. Seven ewes, Simon thought they might buy.

Fog descended. Was that really the right way to go about buying sheep? Where had Dai Sheep got them from? Had Simon and Caleb inspected them properly? Yet, if they really were that good, perhaps we too should have a few extra. Our lambs were fattening well and soon would go to market. Grass was plentiful . . . and memory short. The ministry man had said four to the acre, which meant, by goodness, eighty sheep. It would be safe enough, surely, to have another six for ourselves? Plus Simon and Caleb's new seven, plus the original sixteen and Sam-Ram and his successor. Thirty; surely that would be about right?

We hesitated and procrastinated as nowadays we did in all things, befogged, leaving to our two young men – who thought they could see in the thickest fog – instant decision. The farm year, our first as sheep farmers, was beginning to die, but we still had a few weeks remaining before the new farm year rolled around, and we had one remaining job – the selling of lambs.

Whilst lime and phosphate had begun to transform the vegetable plot far beyond expectation, and the first trees had begun to live even in that conglomerate, it was impossible not to see that the darlin' little lambs of March had become the rumbustious hooligan sheep of late August. Faithful old ewes had hourly to withstand charges that lifted them off the ground as their bovver-boy twins greedily demanded more from the udder. Tail-swinging, lusty lambs bucked and raked about the hillside in magnificent joie-de-vivre, formed themselves into skinhead and mod gangs of adolescent frivolity, went off in pairs to dispute dominance in butting tournaments of frightful ferocity and, worst of all, began to mount each other in lusty practice for the coming rut. They must go; something just had to be done and for us, as usual, dear John Jones was the only man to do it. Once again we screamed for his help.

That August Bank Holiday Monday, the lambs whose arrival into the world only months before had brought such pleasure, now were to grant us, if not pleasure, then certainly relief by their departure. Knowing full well by now of our lunatic inadequacy, John Jones, Uncle Dai and the lorry-driver formed, with the glorious Juno and the driver's dog, a nononsense round-up gang which forthwith bustled and drove the total flock, except for Lucy, straight down off Dolgwili, along the railway track and the footpath straight into the waiting lorry. Being sheep-farmers of such crass timidity that we dare not venture to the market even to see our own sheep sold, we sat waiting until the lorry brought all our ewes back alone, and later until John brought news of the sale. One wholesale butcher had bought all our lambs for an average of ten pounds forty pence each, their average grading was thirty-eight with the best at forty-four, that being, he explained, forty-four pounds dead-weight, yes yes, approximately half the live-weight.

We had, John avowed in a surprise he did not bother to hide as he scratched his head, we had done very well . . . considering. Yes, yes. Very good.

No sooner did the cheque come to mark our year's end, and with us taking a smug breather from the year's shepherding, than we were confronted with the new sheep year. Suddenly Simon and Caleb were driving their own seven new ewes, bought from Dai Sheep, on to Dolgwili.

"They're a damned funny looking lot, aren't they?" I went out and croaked critically in greeting, peering at his pride-and-joy seven. "What are they? Not sheep, are they?"

"Nice and big," Simon pointed out whilst Caleb smiled wanly at his side. He was a nice young man; I was enjoying myself.

"Nice and grotesque, you mean."

"Big ewes have big lambs," he argued, granting me no smile.

"Where did they come from? Africa?"

"Dai Sheep."

"But that one there is a white rhino, isn't it? What do you feed them on? Savannah?"

"Better get Sam in," Simon said sternly, tired of my jeering.

The old gent had just limped over to sniff around his strange new harem and let it show straightaway that he thought no better of them than I did myself. His limp immediately became more painful as the full realisation dawned that he would soon have to cope with the demands of these scruffy old bags of sheep. Soon I called him over for nuts and sympathy. We installed him in the intensive care unit of the stable, settled him comfortably on lashings of bright clean straw from our only bale, with the genuine intention this final time of getting his poor feet good and sound.

All through September all of us waited on him hand and hoof. He lorded it there whilst we fed him cakes and ales, told him ribald tales and whilst Simon earnestly pedicured him almost daily. On Michaelmas Day he was in fine fettle at last. We opened the door, he baa-ed his gratitude, tipped us and walked

73

tall up the hill. As if to inaugurate the new sheep-year correctly, he forthwith tupped an ewe, choosing what seemed to me through the binoculars the very blousiest of the Simon-Caleb chorus for his nuptials.

10   It was a winter of blackness

The wintry nights were filled with unease. Disquiet breezed through the mind at bedtime regularly at the thought of our ponies and sheep out there somewhere on the hill. Over-sensitivity had set in; they were hardy enough and Dolgwili's conditions were cosy compared with many.

This night of December was pitch black. Wind rattled the old window as I stooped to gaze uselessly out into the blackness and from the kitchen below came the sweet chime of the kitchen clock for eleven o'clock.

"He's later than ever tonight," his mother grumbled sleepily from our bed.

In a minute she added, "Is it raining?"

"Not at the moment."

It did not matter either way. Dolgwili was already sodden. John Jones could not remember a wetter summer, but autumn had come worse and now December was a continuous vast wet misery. Day after day the gales blew in from the Atlantic, huge drifts of solid rain sailed past the windows, the leafless trees clung black and desperate to the wet crags and all the hours were grey or worse. The sheep huddled under the hedges, the ponies waited all day in the shelter. The thirty yards of Sheep-dip from gate to shelter had become a morass of churned mud, whilst up at the animal crossroads by the Spout it was even worse.

All was quiet now, as I stood watching and waiting, except for an occasional car along the road, but in the morning donkey, ponies and the thirty sheep would be impatiently calling to us for feed from first light. Ever since mid-November we had been feeding more and more nuts. The grass was all gone, growth had stopped.

Where was the damned boy?

"I should come to bed."

A car came slowly down the road from the village. "I think this will be him now." I watched the car lights turn in off the road towards Dolgwili. "Yes, it's him. No wonder he can't get up in the morning."

I closed the curtains, took off my dressing gown, climbed into bed and held her hand. All was well, with family and farm, everything settled for the night.

"Goodnight, my love."

"Goodnight, darling."

Minutes later the donkey brayed, announcing morning, mocking night. Ill-temper seethed as usual whilst I groped a way into the new day, thirsty for my temper-restoring cups of tea.

Simon was late coming down to breakfast. I cast my day-greeting over my shoulder to him and heard his mother exclaim, "What on earth has happened to you?"

I turned. His eye was hugely swollen and black. He held his hand awkwardly. It was scratched, wounded, swollen. He looked sore everywhere.

"What happened?"

"They put the boot in."

"Who did?"

He smiled wanly, crookedly.

"When? Tell us, for goodness sake."

They had been waiting for him. They came, three of them, attacking him just as he was getting out of his car, just as I would have been pulling the curtains close.

He had said, "Hullo, boys, what do you want?" and the three of them had jumped on him, punching, pulling him down and then kicking him into near senselessness with their hob-nailed boots before stealing his wallet and dumping him back into his old Renault.

"I remember now, thinking how long it was before you came in."

He had come to in the car, crawled home and into bed to sleep it off.

"Did you recognise them?"

"They were the fishing boys. The poachers," Simon said,

76

whilst Phyl so tenderly peeled off his shirt to display the shock-ingly bruised and wounded pale skin of his long thin back.

Ah, father and son, I sighed as I dialled the police, remem-bering the time last summer when I dialled the police, this one the echo of the other.

Last June a small lilac-coloured van had arrived on the place by the river where anglers park their cars. Two youngsters had got out flaunting fishing-rods and addressed me as 'Guv' as we passed, I to shop, they to fish. Evidently they were then on reconnaissance, for a couple of hours later they had exchanged the fragile and untrustworthy rod for the lethal limebag.

The Gwili, everyone said, was a poacher's river. Tree cover was good along much of its secretive length and access along the old railway was easy. From Haydn – another elderly to whom Simon had attached himself most admiringly and who indeed had taught Simon almost all he knew about the river and its fishing – had come endless poaching tales from the past, told with gusto, gesturing arms and great laughter, and all great fun. All right, the poacher who tickled the single trout, or gored or netted salmon for his needy family, this was a folk hero who, remembering the hardness of life in these valleys in the old days, was at least acceptable. Nowadays, however, he had been replaced by the poisoners, greedy men who with lime, or bleach or cyanide could remove not merely salmon and sea-trout but all life in whole stretches of the river that they desecrated. Already two incidents had occurred in our short time at Dol-gwili. The two youngsters with the limesack last June were embarking on the third.

They were, however, so careless in their approach to their foul purpose that they had spilled lime everywhere and already Simon and the hotter-headed Peter had gone after them along the river. I, working separately, found lime spilled on the bridge and inside their Minivan. What to do?

I phoned the police.

Meantime our two goodies had caught up with the baddies at Haydn's pool only seconds after they had dumped the limebag into the river.

"You take that out quick," ordered Peter belligerently.

Well, what did happen exactly? What was said? We had it afterwards only in the incoherent and vague words of Simon and Peter themselves but evidently the threat of one side throwing the other side into the pool led fairly quickly to an armistice whereby, Yes, they would remove the limesack, the poachers agreed and Yes, they would not report the incident, our young gamekeepers agreed. No fish had been harmed, no fish had been taken. The incident had been settled in the Welsh way.

Except that some damned old fool had happened along and messed things up by phoning the police.

"I did phone the police," I told Simon as they arrived back at Dolgwili.

"You did?" Simon exclaimed, sharply staring at me with unusual dismay, although he then said no more in explanation. He walked away and soon the police – an inspector and two constables – arrived. A ton of evidence was strewn around still and the police had no difficulty in charging the two poaching youngsters. They were given the usual nominal sentences – probation and a small fine – and the matter should have ended there. But, of course, it had not. The two poachers, plus one, waited until December and then jumped Simon.

Ah, father, ah, son. Two individuals, living their own lives in a way each thought fit, with their own standpoints, own morals, own background, yet, being father and son, imprisoned together in the cell of kinship. I went out among sheep and sat there wondering about him, about it, about me. He was, at the very least, half the reason we were at Dolgwili at all. Why was it, when you desired to give your children so much, that they were quite unable to accept it from you? Why, when you wanted to make life so right for them, did you yourself make it so wrong? And me, seeing the incriminating lime on the bridge and in the van. Should I have turned the other way? Was I not an individual too? Must I deny myself right action merely because my son was on the riverbank? Ignorant of his negotiations, how could it be wrong to phone the police? Yet having phoned, did I not thereby ensure that the poachers' boot would go in?

A week before Simon's beating, Trudi had – with some difficulty – been delivered of her second child, Ben. After only two days in hospital she had been discharged, crawling back home pale and still weak to try to cope on her own with the new babe and one-year-old Hannah. Peter was away in Devon, leaving

Trudi to lace her child-minding hours with anxiety for him on his return. Would he, too, be greeted with the poachers' boot treatment? So she too became ill.

The police quickly enough identified and charged Simon's assailants, Trudi moved in to be nursed by Phyl, and eventually Peter returned and was never assaulted. The whole affair however had unnerved me. Our Welsh world was suddenly so alien. Rainstorm and wind represented enemies at our door.

So I walked my Dolgwili cloisters, grateful for their solitude and always aware of the approach outside of the Four Horsemen of the Apocalypse.

Old friends died, it was a dark time, indeed, and then Simon was suddenly, even casually, engaged to marry a wench we had not met. Phyl and I began to look at each other with half-wondering, half-apprehensive eyes; did it mean, then, that we should no longer be parents, that just-like-that, after twenty-something years, we two should be on our own?

Then, one wintry day, birds began to collect in the sky; Hitchcock birds. Buzzards, ravens, crows, daily they increased. They circled and wheeled like vultures. Periodically they peeled off with gibberish noise to land out of sight, just over the brow of the hill opposite.

"Sixty-nine ravens today, can you believe?" dramatically I reported to Simon as homewards he plodded his weary way that evening. "And eleven buzzards, all in the sky at the same time."

"It's Blaeneigr," he answered, stopping to stare across the valley. "They say nine beasts have died up there."

"Died? Of what?"

"Starvation," he said in that tone which suggested that I should have known, that I was so out of touch. "They've used up all their hay and they cannot afford to buy more at these prices. Chap at work today told me that three thousand bales of hay have just fetched fourteen thousand pounds in Central Wales."

At least one young local farmer had killed himself rather than stand and watch his animals die. Friesian calves in the market sold for fifty pence, or not at all. What was happening to the world?

Even the weather grew more spitefully hostile. Rain fell mercilessly, overflowing the ditches and bloating our little river to a turgid monster which roared and devoured whole shaking trees in its brown maw. Boots on the lawn squelched in surface water. Earthworms drowned. Rust entered our souls.

"Do you remember the black winter of 'Seventy-four?" I shall croak in years to come. "Whole gates rusted away, just dissolved completely between November and February and got washed down the ditches. Quite young men had to be treated for sogginess and old Mrs. Evans the Post could not stop her fingers from dripping, even in bed. Aye . . . I remember the mud near the barn was so deep we lost three of our best ewes there at Christmas and they didn't turn up again till the dry came in April. Aye . . . Aged a lot that year, I did."

Learnt a lot too, that year, he did; found many answers that year, he did. Turned his back away from the world of men and their affairs, he did, turned in on his damned old Dolgwili and its animals more and more, a middle-aged Noah at rest, even if not at peace, on his Ararat.

Dolgwili was a hill cut off from the road by railway and river, a mile from the Welsh-speaking village from whom we were cut off by language anyway. We had, at our coming here, stepped out of our own context, out of our pecking order, out of our own territory. We had no prestige, no niche, no history, no friends. So Dolgwili was all our world; its comfort, its orderliness, its beauty and well-being became our total concern. Beyond was chaos and nonsense. Because we could do nothing about chaos and nonsense out there, everything on Dolgwili became that much more significant day by day.

It sounds as though the banner of self-sufficiency was being unfurled on our hill. Not so. I had been seduced by that gorgeous hussy several years before back in Suffolk and that affaire had ended in tears. I would not court her again, however appealing she was, for she could lead you into a wilderness, or drudgery, or crankiness, and although I waved to her from time to time and heard the echo of her siren-song, I was scarcely tempted to go with her. I did admittedly pass the time of day with her more than once, for our wood was full of winter logs, our pastures full of lambs and wool and even leather and milk, our river full of clean water and good fish, and presumably our wind full of power, our sun full of heat and light, our hedgerows of wine, our garden of vegetables, our orchard of fruit. But when she coaxed me to go all the way with her, this self-sufficiency minx, I bowed out with a doubting smile.

Simon had not yet gone his way completely, but every now and then we would pull up short, blink and think, "Simon? Getting married? But . . . But, dammit, he's only just left school, almost. What happened to the years? And anyway, if any sort of wedding day ever arrived for Simon, surely he would marry a twelve-bore, or, say, a springer of superior breeding?" And anyway, he had probably got it wrong. His life was compounded of error and misjudgment. It was probably two other people who were getting married.

In November he had announced with frightening profundity that Sam-Ram was so bad on his feet that he could not mount his ewes any more without a 'bunk-up', that the whole of next spring's lamb crop was in grave doubt, that he, Simon, fortunately could borrow a fine young ram from old Dai Sheep or from the next village. Simon could enter the kitchen, with Caleb in support, and make such an announcement with complete confidence and authority and we would believe him utterly. Black Collar needed worming again, he would inform us, or Nice Fleece was scouring, or Blue-Bum had just taken off and flown across the river, and we would look up as seriously as he was in his informing, and we would nod and accept his announcement as gospel-truth.

It was the fog again. Welsh fog. Welsh fog. It gets into everything. It dilutes the truth, and hides it.

So, when our sprog shepherds had pronounced in November on the sexual difficulties of our elderly ram, we had hardly expected to be recording in December that our reputedly moribund ram now looked 'fat and fit' and had beyond doubt once more marked every one of his ewes. The marking was by a blue or red crayon fixed to the old gent's braces, and it recorded for us Sam-Ram's nightly prowess and thereby the lambing date of that ewe. The raddle was Simon's introduction and it was the most important single development and aid to management we had seen on Dolgwili. It was, this raddle, more than that even. It was a searchlight, our first real searchlight to penetrate the fog of ignorance and inexperience which so far had surrounded our sheep-farming.

Sam-Ram happily roamed the hillside with his assorted

wives, carted them off to our sheltering wood whenever the storms blew in, and each morning brought them down to be handfed their half pound of nuts each. When grazing became poorer still and the weather worse he led them through into the better haven of Morgan's Wood, finding entrance not through our spanking new fencing, erected by Simon-Caleb with surprising and admirable efficiency, but through one small unfenced section where the hedging had been deemed strong enough without wire. The fencing had cost around one hundred pounds and for that about two-thirds of that eastern boundary had been made safe, but we had taken fright then at the cost and so once again an ill-afforded winter of my life, and of Phyl's, was to be spent exhaustingly in Morgan's Wood, calling 'tot, tot, tot,' and leading back the flock which so unwisely had been enlarged.

The six new Speckleface ewes turned out to be an uncooperative lot, wilder than our original sixteen and even on the hungriest days unwilling to come to hand for grub. There simply was no rapport between them and us and inevitably we named them the Wild Lot. The seven fresh ewes which Simon-Caleb bought expensively for themselves from dear old Dai Sheep were, we decided, more than somewhat shopsoiled and definitely rhino-shaped. We privately named them the Jumble Lot and considered that at seventeen pounds per ewe Simon's admiration for eccentric old country characters was a somewhat expensive aberration.

He would learn, we chuckled to ourselves as we began to face up to a Dolgwili which soon would know our son no longer. And really Dolgwili hardly knew him anyway, so short seemed the time he had been there.

"Well? What do you think of it?" I had asked him the very first time we had strolled about the place, inspecting the prospects. We had just had an encounter with a salmon, and he had smiled and, in an expansive mood engendered by the fish, had answered at what for him might be termed effusively.

"Couldn't be better," he had said.

The river had been very low that first day. It was quite new to us, and although the estate agent had assured us that it contained fish, both of us as we wandered and peered about us, doubted it with all the doubts allowable to strangers on a river armed with only an estate-agent's assurance. "A quarter-mile of valuable fishing" we were inspecting, stepping from boulder to boulder in the August river, Simon ahead of me and carrying almost uselessly a roach rod with the weights, red-and-white float, and even an old dried-up worm still hanging from it, just as it had hurriedly been thrown into the car for that first weekend visit to our future home. As I had stood on a rock wondering at the beauty of this little river's colour as it tinkled through the sunlit cathedral of overhanging trees, I had taken idly to examining the play of light on water, on weed and stone, as it all flowed past in endless dancing movement on that sweet day of wonder. And there before my very eyes, resting in the current, lay a great fish, a fish of unbelievable size and strength. I had gazed at it for a full two minutes before I lifted my face and whistled to Simon. He looked up from downstream and I beckoned him back with a single cautious movement of my head.

He idled his way back. I kept gazing down at the fish. Only the wavering tail proclaimed it to be alive. Otherwise it was a stuffed fish, a plastic fish, a dead fish; it could not be a real live salmon, surely.

"Try your worm," I whispered after he had joined me stealthily and gawped at it.

Simon gently lowered the unappetising worm in front of the great fish's head, manoeuvred it, withdrew it breathlessly, cast it perfectly, quietly. The moments were tight; disbelief encompassed us.

"It's no use. It's diseased, dying, it must be," I judged.

Minute by minute we stood staring, him casting, me watching. It was no more than two yards from the rock on which we stood. Time after time he withdrew the worm and recast it, time after time the salmon ignored it, eight, nine, ten, twelve times, our caution evaporating, our words growing louder, fifteen times, sixteen, the worm bouncing now on the noble head of that fool of a fish, until suddenly it struck, and was gone with the worm and the line.

Simon held up his puny broken line and we smiled at each other in wonderment that we should have been allowed here in the presence of such a fish.

"Couldn't be better," he had said. Superlative praise.

No, it could not have been better then. It was quite perfect then, for it had no reality. It was all dreams. Life then was all in the future tense, without river-spoilers, slugs, frost, foot-rot, without even an early marriage.

It is no good pretending that at that Dolgwili dream-stage I did not think of it all as being for the children, or even grandchildren, as much as for ourselves. Always they were in our mind. Here they could grow in the ideal of the simple pastoral life, and together, we dreamt, we would make of it a place of order and beauty. The cottage would be transformed into a warm and lovely home, the garden a delight, the farm an unambitious but tidy place of neat hedges and splendid animals. With Simon's agricultural training and Peter's forestry experience we could lead a pleasant and totally secure existence against the threatened days of the Four Horsemen, even unto the days of grandchildren, all of us could grow in the pure air and splendid surroundings of this unspoilt countryside.

I can recall the exact spot and moment at which the dream fractured, the moment of truth.

I came up the slope from the river with the saluki and the aging boxer, stopping for no reason under the giant boughs of the fallen elm. I saw all about me the things which we had between us not accomplished, the failures which so tormented my sense of rightness; the metal field gates, already rusting, instead of the timber gates I had desired, the remaining rubbish still not covered after the departure of the builders, the hairy lawn which Simon was supposed to have cut, Hannah's swing never correctly hung, now never used but left to moulder there, the fence along the garden hedgerow which Simon-Caleb had erected the wrong way up so that lambs could get through the large mesh instead of being excluded by the small mesh, the lack of a store of logs ready for winter, all my unfinished garden, a spade left out. We had done those things we ought not to have done; we had left undone those things which we ought to have done. There was no health in us, O Lord.

Often enough outwardly, and lately inwardly, I had raged against our young folk, against their selfishness, their thoughtlessness, their slovenliness, their lack of judgment, but now as I stood under that black elm with my quiet dogs a moment of realisation came to me, a final facing-up to a fact that had been long growing in my mind, and I said aloud, "By God, they don't give a damn. Not a damn, any of them."

From that moment a fine peace flooded through my mind. The inner rages against our children were done with; I was my own man. Phyl and I were done with the next generation, they were done with us. And it was right that it was so. Until now I had been trying to press *my* Dolgwili dream upon our children, wanting them to live *my* life, but of course the biological urge required them to live their own.

So I was free. To hell with them. Henceforth I would do it all myself, clean my own drains, grow the vegetables myself, treat the foot-rot myself, make my own Dolgwili, or rather make *our* Dolgwili, for all that pleased Phyl pleased me too.

We seemed to have taken off our wellingtons and our rain-coats hardly at all that winter. Day followed day of continuous soaking rain, churned mud made a misery of walking, we had no hay, no dry logs, little peace and the animals seemed perpetually hungry. The ponies stood about in wet unhappiness, looking scraggy on their meagre diet of ash bark, brambles, apple twigs and grass soup, whilst the sheep complained loudly of their hunger every time we showed face outside the door. What capital investment we had started with that winter had almost completely disappeared, gone down the drain along with everything else. It was a winter of blackness and wetness and worry.

By mid-February we could almost feel bankruptcy in our bones. Virtually our only visible asset was the fair bite of green pasture on the lower slope of the garden acre, and by now, just as last year, the sheep were lining up with their hungry eyes on this too. Who, after all, was I to fight against centuries of sheep tradition? It was the measure of my retreat since last year that the fence I now erected was a semi-permanent affair of reliable strength and proud straightness, of solid larch posts and sheep wire, right across the acre. We opened the gate from Sheepdip and the ewes poured in, wolfing the good grass in their ragged sheep's clothing morning after morning, heavy in lamb, clinging like us to each slow day, surviving, surviving.

11   They take you by the heart and squeeze and squeeze

Each morning, as the tide of sleep ebbs, one stoops to peer out through the small windows of bedroom, bathroom, landing, sitting-room and kitchen in turn. The windows are small, of a size perhaps dictated by the nature of the local stone of the sturdy walls, and each window successively frames an outlook up or down Dolgwili's compass points. One looks out without eagerness; hardly a morning arrives but what apprehension comes too. What of today? Have the sheep survived the dreadful night?

The frantic jackdaws clack and swoop in black gangs from the chimney to the doomed elm, whilst down the road the toy cars rush and rush and rush. Across the river old Dai Sheep's ewes are up under the hedge for shelter again. Our own ewes are just descending from their night place across into Lower Rough, and somewhere in Sheepdip the donkey brays her demand for an early crust. The roar of the wintry river is so constant one does not hear it; it is like the roar of your own bloodstream in your ears. Chaffinches brightly examine the terrace paving for crumbs from last night's dogs' dinner. It starts to rain again.

And through the gate of Sheepdip into Garden Paddock a tall young man emerges with a gun under his arm, a white terrier at his heel and a lamb under his arm. He holds the iron gate open, commands the lively dog to sit and encourages the ewe through.

"Simon's got a lamb," you announce with surprise.

"Oh, good. That will be Gentle's." Phyl comes to look through the window. "She was first to lamb last year too."

The ewe begins to graze my tender lawn grass with relish and the lamb noses around her belly until it finds the udder. The young man stands watching for a minute and then he and the dog come up into the kitchen. Breakfast is cooking.

"Where did she lamb?" Mum asks from over the frying pan.

"Usual place."

"What is it?"

"Tup. Do you want me to bring old Dai to castrate it for you?"

She looks at me in foggy consternation. I shrug at her in foggy consternation. "Is that good?" I venture.

"Should we? Is it better?" Mum asks him.

Simon smiles his knowing smile. "Dai always does his *own* lambs," he says, as if he is quoting the gospel.

"Why?"

Simon is silent. The all-encompassing fog has come down again. None of us really knows what he is doing, nor why; we farm in a fog of unknowing. Nevertheless our morning faces wear the slightest of pleasurable smiles, and Simon and I are no sooner into our eggs and bacon but Phyl has finished her yoghourt and is into wellies and dufflecoat and reaching for Grandpa Kirby's famous walking-stick. It is the beginning of March, the beginning of a new lambing season.

Blackface Longtail had lambed on Sunday, but because one of her twins seemed a bit weakly we kept her in the stable, together with the ewe which had been injected by the vet for leg trouble. On Wednesday morning when we opened up, Leg Trouble had had twins too. There's lovely . . . Four lambs, on clean straw, what could be nicer?

It was a fair day, comparatively, both ewes were ravenous even after a good breakfast of nuts, and although there was a rather vague debate at the stable door on the benefits of keeping in against those of letting out, we carried the lambs across the quagmire in front of the stable and they went happily off across Garden Paddock to rejoin the flock. Phyl thought that the lamb still looked poorly and later indeed we noticed it was scouring, but when later the same day Black Collar lambed down successfully, although with only a single, we began to puff ourselves up a bit too much on the start we had made again. Each time we met old Sam-Ram we saluted him by offering a cob, which he greedily grabbed, together with the unbegrudged two joints of our middle fingers. Good boy, good boy, we would say with that knowing leer that seems to be the correct recognition of virility.

He had a most noble baa, did Sam. It was deeply resonant and quite distinctive, and he held his large black head very well as he spoke. The truth was we were half-besotted with the old twit. His fleece never broke as did many of the Speckleface ewes, although it did always look shabby and wrinkled, and of course his poor feet were still never right for more than a week or two at a time, especially with this season's fearful mud. Regularly Caleb and Simon trimmed and sprayed them, a toilet which the old gent accepted with grace before, patched up once more, staggering back up the hill with that air of broken down nobility, to his tatterdemalion harem and his exquisite offspring.

Morning feed-time had now attained a routine whereby I lured the ponies and donkey to behind the shelter with bowls of pony nuts, whilst Phyl called the sheep to be handfed across on Lower Rough. Fleur, boss pony, was almost certainly in foal again, so her rough demands for all the food could end in a series of running battles across the slippery mud between ponies, donkey, me, sheep, lambs and Phyl unless the presentation was most carefully stage-managed. We flinched each time Fleur lashed out with flailing hooves lest a lamb be within range, and Phyl and I were always edgy with each other if we failed to synchronise our feeding time, her on Lower Rough to the east of the garden, me on Sheepdip to the west.

The rush of ewes to Phyl invariably meant that lambs got left behind. Whilst mothers surged around and nosed at Phyl, lambs bleated and scampered here and there continuously in frantic search for their mothers, and only afterwards, when hunger was mostly appeased and it was obvious that breakfast was over, only then did the flock sort itself out, only then could roll-call be taken.

That Thursday was once again the direst of days. Out of the west all the remainder of that dark winter's arsenal of fearsome rainstorms came hurling themselves accompanied by howling dervishes of wind. And the roll-call would not come right. We leaned on the stiff wind and counted and counted till we were blue in the face. The rain lashed into us, ewes lost interest in us. The flock wandered disconsolately off in search of grass that was all of two months away as yet. We had no hay, and there was a limit to the amount of concentrate that safely could be fed. Newborn lips and tongues would be sucking away through those black teats all the remaining nourishment of the ewe herself, draining her of body reserves. And now one lamb was missing; there was no pleasure abroad that morning.

"It's definitely that weaker one of Blackface Longtail's that's missing," Phyl said again, worriedly looking about her towards the Steep.

I recalled the lamb's dirty behind and the mincing trot it needed to keep contact two mornings ago. "We'd better start looking, then."

Phyl went eastwards, I went west. It poured. Where do you look for a lamb in twenty-three acres of winter? Once again I prowled every hedgerow of every pasture, all through that damned wood, down the railway bank, everywhere. The rain took to pouring horizontally instead of normally. Thighs soaked first. Hands and face stung into redness, yet soaked cold. You kept your head down and looked sideways. You went slowly. You had never lost a lamb yet. Hope drowned slowly, but you kept on.

Blackface Longtail herself was looking too, deserting the flock to climb the Steep, bleating, standing, looking, waiting. Our own wood started there, impenetrable, solid prickle and thorn. A lamb could easily creep in there, but we could not. The spite within the pouring rain increased. I made a token search into the less dense part of the wood, wetness trickling into the warmth of my back now and down into my wellingtons. I trudged on spiritless and as I came out four damned ewes had the gall to go pushing through the boundary hedge out on to Thomas' farm. If they were doing that, then perhaps the lost lamb had been stranded up there earlier this morning. So I pushed through after them, searched around in vain, brought them back and plugged the hole as well as I could. I went home.

It was eleven-forty-five. Phyl had just changed into dry clothes. We had coffee and some sadness there in the kitchen. We had lost a lamb.

We did it all again in the afternoon. Up on the Steep the villainous rain lashed us mercilessly as we leant on the wind and searched hopelessly on. It had gone. It would have been weak from its scouring and from the weather; a fox could have taken it. We were very angry with ourselves, for we had been disgracefully negligent.

At dinnertime, home from work, Simon too was strangely upset by the loss. He lectured us a bit on the issue, a turnabout process which took us by surprise and accorded me some private pleasure. He further instigated a new rule, which pleased me even more, that henceforward all new lambings should be brought into the stable without exception for the first three nights.

We were learning. We were not farmers, of course, we had always excused ourselves. Oh yes, we had read the books, and oh yes, we ought to have hay, of course. We knew what ought to be done, but how could you do all that on twenty-three acres of hill, with poor access, no hay barn, no equipment? You could not take a tractor on these slopes and anyway it was quite absurd to consider anything approaching what we thought of as real farming until the railway was well and truly dead. Each sack of feed, each bale of hay, the shopping, coal, everything still had to be wheelbarrowed from road to cottage. Access for heavy loads via the river ford and railway was denied us all winter anyway because the river was always too high, and although the old milk train no longer rattled and squeaked its way past us, our negotiations with British Rail to buy our vital stretch of access-track rattled and squeaked on about as fast as the slow old train had done. The lines grew rusty, weeds and saplings began to sprout and spread in the grey stone of the trackbed and whatever fine farming ideas we conceived from our reading of books or the farming press, or whatever quite ordinary farming ambitions we nurtured, all of them were thwarted by that fifty-yard stretch of railway no-man's-land. Feet yes, wheels no. We were learning, but we were learning the hardest way of the lot.

Marooned in our village one day, I met old Dai spitting into the river from the bridge, and falling immediately into Simon's part of straight feed to his part of quaint old country character, I had, after explaining how much trouble we had with foot-rot, asked him what his own remedy was. From the image of Dai which Simon had tended to purvey to me, I naturally expected instructions, laboriously and quaintly translated from Dai's natural Welsh into approximate English, that I must sit up three nights at the full of the Lammastide moon to prepare a tincture from oak bark, kingfisher wing and wool from a ram's tail and then apply this to the foot-rotten ewe at dawn of next market day, or similar. But no.

"Well, er . . . " Dai said, clearing his throat once more and wiping his wet nose on the back of his bony hand as he looked away at the distant hills. "Squiss, aye."

I stared at him, fascinated. "Really? What is that then, in English?"

He stared back reproachfully. "Squiss," he shouted, impatient at my stupidity.

I shook my head, furious with myself for losing out on such wise old traditional practice which might be lost for ever should, for example, I obtain an inaccurate translation or should this grand old character topple off the blasted bridge.

"A herb, is it?" I bellowed back at him.

"Naw," he retorted loudly as though I were the foreigner. "Squiss, squiss," and he held up before me his bent aerosol thumb and squirted it. "Get it from the vet. Squiss."

Few answers, I decided, were likely to come from dear old Dai. Nor indeed were many answers likely to come from books or ADAS; they would come only if we searched for reality here on our own blessed ground, with diligence and intelligence and luck.

The evening after Simon's shamefacing lecture, all three of us went to the hill and brought in, after no more than two hours, the five ewes with all their lambs who now qualified for three nights' intensive care under the new Simon regime. Gaining in confidence, Simon pushed us further with instructions that henceforth too we must matchingly mark each ewe and lamb at birth, and as we finally packed up that night, shutting the stable door on ewes bedded safely and cosily down on clean wood-shavings, we all really sighed with satisfaction not merely for the safety of those sheep but more for the lessons we had learnt.

"Phyl and I are tired out each day now, but rather enjoying it all," the farm diary records. "Watched the hunt right up on top of the hill opposite. We saw them shoot one and later heard they had four during the day. Old Dai was right up there with them. Simon says he heard in the pub they have had more than two hundred and fifty foxes in the season."

It was a nagging time. The old winter kept nagging the young spring as regularly as a parent, the ewes kept nagging the hedges and the lambs kept nagging the ewes. It still was not a happy time, for all the arrival of lambs.

The six new Speckleface we had bought from John Jones were always toffee-nosed and wild-eyed, ready to break out. Only one produced a lamb, and her we named Wild One, for no sooner did she see us arrive that she bristled and rose up on her toes to flee. And what a dance she led us at dusk when it came to stabling her and her lamb for those first three nights.

Normally one has only to catch up the lamb and the ewe will anxiously follow straightaway down to the stable. Sometimes one has continually to show the lamb to the ewe, or more probably smell the lamb to the ewe. Neither worked with Wild One. Round and round that damned hill she would run, bleating her head off for her suddenly lost lamb, and even if then contact was re-made between her and the lamb, us holding it and showing it to her as she came closer, then still again she would dash back to where she had originally been with it and we would start the whole frustrating game again. Hours she robbed us of, that Wild One.

Even then we were more fortunate with our Wild Lot than Simon-Caleb with their Jumble Lot. Their Sad One produced a stillborn lamb and was subsequently found to have only half an udder. Another one of theirs gave birth to twins but immediately one was very poorly and had to be rushed to the warmth of the kitchen. It could not stand. If we balanced it carefully on its perfect hooves, it collapsed almost immediately. Phyl took to feeding it glucose and milk from a dropper, named it Polly Flinders and handed over the kitchen to her. Saluki and boxer were banished on Polly Flinders' behalf, we ourselves could settle to nothing else and every few hours was feeding-time again.

Next poor old Blue-Bum was cause for concern. She was a great rhino of an ewe, articulated like a lorry and about the same size. It always appeared strange that she had but four legs underneath all that body. Already heavy with lamb, she had completely isolated herself on the level part at the top of the Steep which I called the Shelf. Two and three times a day Phyl climbed up there to throw Blue-Bum nuts and to have a breathless chat about getting a move on, but the great ewe remained obdurately pregnant day after day. She had achieved her name from the bright beacon of blue shining out from her rump, a beacon to which the eye was immediately attracted every time we surveyed the hill from the garden. First thing each morning this bright blue blob, a magnificent souvenir of Sam-Ram's more-than-wholehearted attentions in the affaire one hundred and forty-odd nights ago, shone out to announce to the waiting world that yet again, Blue-Bum's time had not yet come. No other ewe had ever been so definitely, so colourfully marked; Sam's feet must have been quite perfect that night.

For young reasons or unreasons of their own, Simon and Caleb had around this time taken for themselves a few spare acres close to the Forestry behind the village. They had slaved at hedging and preparing their new territory ready for their seven ewes and lambs and were all set to move their Jumble Lot off Dolgwili at the time when poor old Blue-Bum was sitting out her prolonged pregnancy up on the Shelf. The hilarity engendered by her name, by her state and by abusive encouragement shouted up at her from us leg-weary shepherds below did nothing to help poor Blue-Bum produce her lambs. Already overdue, the ewe seemed to be swelling enormously and never to be moving, more like some sort of queen-bee swollen with swarms of young. In that nagging time, Phyl nagged at Simon-Caleb that they must do something about her, and in turn they climbed up to the Shelf and nagged Blue-Bum to her feet and then down the slope. It was an alarming progress. She could scarcely walk, so heavy with lamb was she. Constantly she collapsed under her own weight and although she seemed well enough in herself, she was just almightily and helplessly pregnant. After each collapse, she would be allowed to rest before our young men levered her upwards and supported her for another yard or two down the one-in-one or one-in-two hillside.

There was no way she would ever travel to the splendid new meadow up by the Forestry until after she had rid herself of that burden of lambs, so beside the house we prepared a brand-new maternity ward no more than yards square, in shelter and of unsheeped grass. Into this cosy little place poor Blue-Bum was at last delivered, there once more to collapse and to await like some immense and comfortable matron her confinement.

The days grew stingy, with thin snow blowing in the wind;

93

one could hardly blame Blue-Bum for non-delivery. We came home from church on Good Friday, however, in one more fierce snowstorm to find that she could hold on to her precious cargo no longer; she had given birth to four lambs. The small cheers we were about to let go into the teeth of that bitter wind for Blue-Bum and gallant old Sam-Ram died as we observed the weakness of two of the lambs. The dying cheers turned to rough admonition very quickly, for our erstwhile shepherds were disgracing their title not fifty yards away. They were hedging, hedging most energetically and indeed so concentratingly that they had given little attention to Blue-Bum. Oh yes, they knew she had lambed, and oh yes, quads.

"But, dammit, these are your lambs, it's your ewe. The weather is bitter and you've got one lamb almost dead and yet you carry on hedging as if it were no concern of yours. What's the matter with you?"

The generations were at each other's throats again. Oh dear . . . They stared, dropped their tools and began to persuade the great old ewe towards the stable, carrying her two stronger lambs whilst Phyl and I moved the two weakly ones into the kitchen alongside Polly Flinders. It took no more than a quarter of an hour's solid warmth from the Aga to resuscitate the first lamb enough for it to be climbing out of its cardboard box to suck at fingers, buttons, anything in its desperate need for milk. The second, which at first we had thought to be doomed, took a couple of hours even to hang on to life, but gradually she too came strong enough to stand and to suck. The Simon-Caleb team, alert almost too late to their responsibilities, rushed about, to the vet (to *their* vet, so much better than *our* vet, of course) for emergency medicine, to a neighbour for clean bedding straw, and no doubt to Dai Sheep for more sheep-wisdom. A day or two later they transported Blue-Bum herself from Dolgwili to their new pasture by the Forestry near the village and the four lambs to other kitchens or other shelters where one by one each lamb died and poor old Blue-Bum, overwhelmed perhaps by the monstrous strain of family life, had given up even the attempt to mother any of them.

The evening following the birth of the quadruplets, when our young men had been alerted to responsibility, I checked as the day died that Blue-Bum was all right for the night. She was indeed nicely snug in the stable, on that new straw with her two stronger lambs, but then I noticed that she had been given no water. I went and drew her a clean bucket of cold water. When I placed it near her head, she arose and drank, and she kept drinking until the bucket was empty.

Moments of wonder arise out of nothing, as if the fog you live in crystallises quite suddenly to a diamond which you hold in your hand and examine. How could such things be? you wonder.

My wonder then was not so much for the way she drank – although never before or since have I seen a sheep drink like that, but then never since have I seen another Blue-Bum or quads – as for the fact that Simon and Caleb had overlooked the water. My wonder at their crassness was intensified, beyond a doubt, by the generation game that we were all engaged in. It was Simon-Caleb versus Phyl-me, outdoing the others, sharpening each other up, competing in the gentlest, quietest way as required by the laws of the game all of us are required to play, by nature or by God or by whatever name we give to the life-force which drives and devours us.

So, as news came that each quad in turn had died, our own foster-mothering in our own kitchen of Polly Flinders became that much more careful and determined. Unable to stand for many days, one day arrived when not only did she balance on the tiled floor by herself, but she further walked and scrabbled into what became an hilarious ballet shuffle around the entire kitchen among chair- and table-legs whilst we laughed and murmured our encouragement. Soon a sunny day arrived and she walked by herself on the lawn, and soon another day when she was strong enough to join the flock.

For poor Simon it did not end there. As if our black winter had not been dire enough, spring came dark with mourning. We came home from a sale one Saturday to hear from Simon that yet another of his seven ewes had died in parturition, and then, almost in the same breath that Greedy, another of his, also had only half an udder. Of the whole seven, not one ewe had been any good. Seven ewes had yielded Simon but three lambs, including the bottle-fed Polly Flinders.

Worse was to come.

Phyl looked out of the kitchen window about noon on a

95

Saturday which Simon had put by especially for hunting. It was April by the calendar, but January still by the weather and we were entertaining a Suffolk farmer and his family to whom Simon was today to show what Welsh hunting was – as opposed, of course, to proper Redcoat hunting.

Phyl said in a puzzled tone, "That's Simon just pulled in. So early?"

Together we watched him toil up through the garden carrying something white. Urchin.

The terrier showed no wound as our son brought him in and gently laid him on his bed.

"What's happened?"

"Got hit by a car."

Urchin could not stand, except on his front legs. From midspine back he was helpless, and although the vet, from whom Simon had just come, thought the paralysis might be temporary, the injury looked dire.

It had been the last meet of the season and Simon and his farmer guest had set off so full of joyous expectation. The cars, loaded with assorted dogs, the guns, the farmers had all been assembling close to the village when Simon's car joined them. Within seconds of letting Urchin out, a Mini had come bowling along and hit the terrier.

Of all the dogs we had ever let ourselves in for, Urchin, even more than his mother Tiger Lil, was the best, a most parfit gentil knight of a dog, fierce in the field, gentle in the hand. They were Jack Russells, Simon's very own, and Urchin was the chosen of that record litter which Tiger Lil had raised so effortlessly.

For days now, helpless in his basket, Urchin whined and cried for the boss each time Simon went dogless to work or to the hill, his eyes hurt and puzzled at this desertion. Daily we tried to imagine some improvement, some alleviation of this damnable paralysis. Unable some days to part, man and dog would share their workdays, Urchin in his basket in that old Renault, or slung across Simon's chest, the dog always watching the man with devotion.

It had been a damnable season.

Old Sam-Boxer this same season was screwed with rheumatism or some such, so that every frosty morning pinned him yelping with pain. His cold bones grew unreliable and his legs gave way from time to time. He needed warmth so much we suspected him of trying to open the Aga oven door to creep inside. And only twelve days before Urchin's accident, Trudi's Jemima Boxer, far gone in age and decay, had been put down. Something, somewhere had it in for us that year. The moment Jemima fell, for all that nothing more was left for her to do but die, in that moment I cried like a baby.

Even Simon, shattered in his mind as Urchin lay shattered in body, and in addition tired out by the frustrations of sheep-keeping both on Dolgwili and his own inadequate patch by the Forestry, announced on a day that must have been within a whisper of spring itself, that he was fed-up with everything. He even took to his bed for a day, as rare an occurrence as had been the passion with which he had declared himself fed-up. I myself did not doubt that it was truly the agony of the doomed terrier which had so dispirited our son. That in turn dispirited us, and although we all hung on for another week or so to the hope engendered by a return of strength to Urchin's spine – suddenly he could wag his tail again – an evening came when Simon announced that he was having Urchin put down tomorrow.

It must have been my distorted imagination that Urchin knew his time had come. Yet he slept soundly all that evening and all the next morning, not struggling about on his two good legs, not seeking our comfort, not whining anxiously for Simon. Outside the rain poured continuously. At some time during the day I went down and dug a small grave next to Tiger Lil's. After worktime, Simon carried his dog down to the vet and then, in privacy and darkness, buried him.

The blasted animals take you by the heart and squeeze and squeeze till your life is no longer your own. They grant you merely their beauty across the running turf, or merely their cold-nosed comfort at the time of loneliness, their quiet nobility and a few lessons on the dignity of patient acceptance which we would do well to learn, but in return for these pittances they demand of you, in the end, all the minutes of all your days, some of your nights, and your tears.

97

12   A new awareness

It was generally conceded, albeit with a faint air of surprise,
that Simon had performed remarkably well at his wedding.
Perhaps nobody really believed that he would leave it absolutely to the last moment to come rushing in from the fields,
throw down his shoulder-stick from which always hung a dozen
cross-legged and newly-gutted rabbits, command his new
springer spaniel to 'Wait' at the church door and then, wiping
his muddy hands on his trouser thighs, hurry down the aisle in
his wellingtons to arrive before the good parson just in time to
murmur 'I do'. And there would undoubtedly have been a little
mild surprise in that marvellously simple little church had the
head of one of the ferrets which our son habitually carried about
inside his shirt actually pushed forward through the shirt-
buttons to sniff at the bride's bouquet during the ring bit, say.
Probably the least anyone expected was that he would have
absently lodged a gun in the church porch ready to leave
afterwards for the more serious business of hunting, from which
this fussy and frivolous business of marriage had kept him. But
no. If he was almost neat – where last had I seen him wearing a
tie? – in that suit borrowed from his brother-in-law, with his
hair brushed, without wellingtons or dogs, only a few of us close
enough to know could tell that he still smelled of hawthorn and
hawk and wild air, that his gaze was past the faces of us present
out to the hill, beyond the wood and over the river, that here he
felt ill-at-ease so bereft of silence.

Still, he said 'Yes' loud and clear for all the old ladies of the
village to hear – and perhaps loud enough even for poor dead
Lizzie, his grandmother, to hear, for she did so love a loud
marital affirmative – and Phyl, at my side, choked a little over
some gritty fragment of time passing which brought a smiling

tear to her eye, and we took snapshots crookedly of little Hannah stealing the show and passed noisily next door to the village pub for cakes and ale and speeches, including one with only half-an-udder from old Dai Sheep, and then our son was gone. And for us, the empty nest . . .

He had gone without scarcely a one of all the fine answers to all the best questions which had taken me at least his lifetime to collect. Back on Dolgwili I was left sitting with a parcel of useless wisdom on my lap. I had offered them confidently to him, but he had smiled that sudden disarming smile at our kitchen door and said,

"I'm just taking the new dog up the hill for training. I'll get them later, as I come back," or some such excuse.

The words of wisdom are on my lap yet, for a son must find his own answers. The span of life seems to be constructed with exquisite irony. By the time experience grants answers to the searcher, it is too late. The searcher himself is too old to benefit from them; the son of the searcher is too young and too busy searching for his own answers.

Becoming therefore only a sort of self-employed guru, with now both our children off our hands and our grandchildren on our laps only until they grew tiresome, I found all things changed.

The sense of release was marvellous. We were liberated once more, just as we, as a family, had been liberated when first we had fled our known East for this unknown West. Then Wales had liberated us from habit, from known ways, from that niche we had long hewn for ourselves in that small country town of our birth. In Wales we were unknown. Nothing was expected of us, nothing demanded. That liberation by Wales of the family had been very real; now this liberation by Simon of us two was just as real.

Slowly we looked about us with new interest, with a new awareness, seeing with new eyes not merely that our enlarged flock of thirty ewes put to the ram last spring had yielded a mere twenty-one lambs reared, not merely that poor old Sam-Ram was footsore still, that our pastures were poor and our fences worse, but seeing it all against the fact that we were alone and very middle-aged. Really, we saw, Dolgwili was a nonsense. What on earth did we think we were doing here at this mad garden-making and sheep-keeping, flatlanders and foreigners, now that Simon was gone? What indeed was the financial sense, all this walking, watching and worrying when the total income from it all was exactly £241 for the year? Why did we do it?

We did not know. May be it was our Everest. It was here, it was ours. It demanded of us our care and we could do nothing else but respond.

Out from old diaries still fall foolish lists headed 'Jobs To Be Done', or 'Budget for Improvements'. You pick them up and read them, and how they mock you. They say:

> Trim all hedges.
> New post and rail around paddock.
> Plant beech hedge.
> Make rose border.
> Invest in goats (must be Anglo-Nubian).
> Make pond for duck-breeding (ornamental only).
> Make stiles.
> Buy donkeycart.
> Clear Steep of gorse and hawthorn.
> Paint all ironwork.

Their mockery comes because tonight's list, were it made, would be little different; each time a list gets made it is one more wish to bring more beauty, more order, to Dolgwili. I came to suspect this was what all this Dolgwili nonsense was really about; it was a sort of creative urge, a need for beauty. We had stumbled upon a canvas spread across twenty-three acres, and some damfool quirk in our nature required of us both that we paint it beautiful with golden bantams and roses, with shape of tree and line of hedge, with animals as a focal point. Our vague aesthetic sense was the source of all our effort, and all our discontent – for disorder was rampant still.

If Simon's departure – geographically only a limited departure, for he had found a cottage in the village – had granted us a new awareness of sheep, work and farm, it granted me a more penetrating awareness of myself as person. Stripped suddenly of my fatherhood, I began to see myself as a man. It was like being stripped of a worn-out suit of clothes, to stand naked before a mirror and survey a different being. It was like seeing the back of your head, or hearing your own voice for the first time. After all the years raking through the motives and behaviour, the thoughts and deeds of so many men, suddenly I had come face to face with the searcher. You mean, that is *me*? Just as repugnant as that first sight of the back of my head and the sound of my voice, I now found so many other chunks of myself to be. But more than anything in this new book of revelations was the overwhelming importance of visual beauty to all my ways, past and present.

Woman, saluki, flowers and foliage, football, ponies, fish, words, clouds, flying, whatever part of life I had ever had to consider, visual beauty had always been the fundamental criterion. She could not sing or cook, she was not rich, she did not even like cats, but the beauty of her young face I could hardly glimpse without shaking and could hardly dare to kiss. The saluki was accident-prone, difficult to manage, independent and a true prima donna of a bitch, but her chocolate, cream and white feathering and her bounding gallop across the hill took my breath away. Same with ponies; they were only on Dolgwili at all because they were so beautiful.

Timidity, or a retiring nature, too, now showed up within that man so suddenly displayed there before me. It astonished me. It had been there all the time, but only now did I perceive it. Concorde, six hundred ewes, the city's morning rush-hour, the ocean, before all such might I shuddered and turned away. At Simon's wedding an agricultural man, who had viewed our Steep, assured me cockily that he could take a tractor up its partly one-in-one slope if ever I had need of such a man. I quaked at the very idea, averted my eyes in dread and assured him almost angrily that never would I have such a need. Fairly small was beautiful; timidity and beauty were cousins, perhaps. So I saw that the same timidity which had prevented me from enjoyably stalling and spinning those old Stearman biplanes in wartime was the same timidity which had sent me running to old John Jones every time I wanted anything done with sheep. That poor old ill-used man must have shaken his head a dozen times in amused puzzlement as he put down the phone after each of my timid requests for 'Did he think he could take our lambs to market soon?' or, 'Could he get the shearers to come soon please?' or could he get a new ram for us, or another few ewes, and so on.

"Yes, yes. Well, why can't he do it his blessed self, you see? Aye, aye. There we are then," I could imagine him saying cheerfully enough as once again he had to set about helping that poor fist of an English twit over at Dolgwili, yes, yes.

Well, there we are then, it would all be different now, yes, yes; do it ourselves, we should, you see . . .

The Happy Wanderers could go, for a start. Blackie Longtail, Bessie Bunter, Disappointment, Greenrib, all the Wild Bunch, and all Simon's duds, any that had borne no lamb, any that were not good mothers. We would start again with a nucleus of manageable, thrifty ewes; we would try to become proper farmers.

"And what about old Sam-Ram then?" somebody bravely asked.

Ever since Sam had arrived on Dolgwili somebody had been sticking their head up to say, "What about Sam-Ram?" Ever since he had come, he had been going. Always at such times when no decision was urgent, he was going. Always when decision was needed, how could we get rid of him?

"After all, there's nothing wrong with his virility," Phyl would point out. "He got Blue-Bum with quads and Spotty with triplets and plenty with twins," and despite all his tattiness I would agree all too readily, forgetting all too easily Sam's inability to chase and hold the Wild Lot who had repaid their winter's keep with not a single lamb.

"All right, we'll have a real good go at his feet ourselves and see how he copes come September. We can always ring old John Jones . . . I mean, we can always go out and buy a new one if he cannot cope."

The Shepherd's Calendar was just starting, the foundation month for the whole year. Just as the book instructed, we gathered into the paddock near the shelter our whole flock for examination of feet, teeth and udders, the eleven old ewes of good character, plus Simon's Greedy and Sad One, temporarily reprieved along with Sam-Ram, plus little Polly Flinders, plus one tup lamb retained vaguely with the idea that he might replace Sam. The Wild Bunch, for whom we had paid John

Jones eleven pounds a head a year ago, had now gone for sixteen pounds thirty each, eight Wanderers plus four of Simon's no-gooders had fetched thirteen pounds each. The Simon-Caleb venture with their own sheep and own pasture up near the village had been a debacle although, as we laughed up our sleeve, they had gained a wealth of experience.

The Simon-Caleb treatment of Sam-Ram's feet had by this time attained the status of ritual. I did not doubt that they had learnt at the feet of the High Priest himself, Dai Sheep, for the young men kept so habitually to the dance and wrestle, the throw, the grunt, the hack, the squish and finally the moan, that they worked as if they were truly the fine inheritors of a great country tradition, as if the routine was quite unchangeable. They worked grimly and tended rather to relish the awfulness of the job, sniffing at every offensive hoof and uttering a nose-wrinkling 'Pough' as if never in the sheep-world before had so nasty a case of foot-rot ever happened. Their chosen tool, too, was always a rather harmless-looking penknife which they never sterilised between sheep, a process which however often Phyl and I had recommended it to our young men, they seemed quite unable or unwilling to try. Such was our generation gap then that they wanted nothing to do with our recommenda-tions, but now, well, we could please ourselves, and I certainly did not favour that wrestle-throw-grunt-hack-squish-moan stuff. The moan they ended with each time was on behalf of their aching backs after they had stood bending over their sheep for all the minutes that it took to treat each foot in turn. Truly they were entitled to it, that moan, for it was harder work than I could possibly contemplate.

Instead, at Simon's departure, we armed ourselves with con-siderable hardware. He and I, at a time when we had shared the chore of lawn-cutting, had once costed our endless trailing up and down and across and round behind that rotary cutter. We found we had walked thirteen and three-quarter miles give or take a likely mathematical mistake or two, and that at that rate I calculated that I might have about another fifteen more months to live before I fell knackered. Therefore, despite the hole in our bucket through which all our lovely loot was so desperately leaking away, we bought a small ride-on mower which almost immediately became a life-saver to me and indeed the garden too.

Then, to make it even remotely possible for me to tackle the foot-rot problem alone, we armed ourselves with a more than sturdy sheep-cradle designed to hold still the most frantic sheep – if only we could get the sheep encradled – with a specially-designed foot-rot bath through which the sheep would walk and thus treat themselves – if only we could get the sheep into it – and real hoof-secateurs to replace those harmless-looking little penknives.

To the new awareness which had come as Simon had gone, was added a sort of euphoric second-honeymoon splendour at this stage of our idiotic farming. We were doing our own daft thing in quite our own daft way, but then was it not far wiser to put the inflated money into such happy-making things as foot-rot baths than to keep it in the official leaky bucket.

Sam-Ram, it seemed to us, was particularly understanding when it came to cradling him. He always was rather different. Quite often he favoured the company of the donkey rather than of sheep, as indeed I was inclined to rather than of man. For days on end he would graze alongside the ponies, and he always was more civilised than any other sheep we had on the place. We never doubted he was of sheep-aristocracy. In the cradle he sat gawky and helpless, but always quietly puffing away at his cigar whilst I busied myself with apologetic small-talk and the nifty clipping of his flabby, smelly feet with my fine new hoof-secateurs. After the pedicure, we booted him out in small polythene bags containing formalin solution, held on at his ankles with rubber-bands, we stabled him most civilly, fed him up sensibly and nursed him all through September. We did not say so, but we each knew by now of our own unwisdom in keeping him. We acknowledged that we suffered from loyalty, even overdone loyalty. Perversely we remained loyal even to our own loyalty, for it seemed to us that loyalty was too often cast aside together with too many other out-of-fashion valuables. We had always stuck by our animals; we would stick by Sam.

He was perfectly content to sit all day in the stable and be waited on while his feet steadily improved. On 8th October we invited him out. He tried out his legs like a footballer after treatment, and then trotted up the hill in his fairly smart, fairly new shoes and began tupping every ewe in sight.

It was a splendid autumn; it began to atone for the black winter and the grey spring.

13 The mark of the tiger was upon him

Outside, the wind grew outrageous, a devil blown in off the Atlantic to besiege the farm, to threaten calamity. Our old windows rattled with fear, blue woodsmoke puffed back down our chimney and outside a dozen dislodged objects danced and bowled about the terrace. The rage of the wind blew into the mind, destroying the evening comfort, demanding of me that I dress and go out again to check that all was intact.

Beside the cottage the four towering conifers were being ravaged by the screaming wind, producing a continuous roar as fearsome as the sea's roar. Torchlight showed their tops bowing and threshing before the wind and my young shrub border threatened to take off en masse. My fear, however, was all for the shelter across in Sheepdip, that the wind would get under the asbestos sheet roof and lift it. The strong beam from the torch shot across the blackness and as soon as it illuminated the four-square intactness of the simple three-bay structure it also illuminated a single pair of yellow eyes. Sam-Ram. He sat there completely at his ease in the end bay and I immediately switched off with brief thoughts of gratitude to whatever gods were in charge of us at Dolgwili that night.

Lately Sam had taken up with the ponies again. He had completed his tupping duties, had his raddle removed, and then discovered that there were good pickings for him each morning and evening on the pony parade. He did limp slightly still, but not sufficiently to stop him from rushing one or other bowl of pony nuts quite ferociously, a game which sent the ponies tossing angrily from bowl to bowl and a game which the dun Syndod, herself now in foal, always lost. During the winter storms, ponies, donkey and ram spent more and more time in the comparative hospitality of the shelter and indeed this was in my mind this evening. However, the ponies never stayed all night there, the donkey, I suspected, sometimes did, and evidently Sam quite often did.

Anyway, all was well tonight. Before Simon's departure, we had weighted down the shelter's roof with half-a-dozen heavy timbers taken from Dolgwili's original barn and I was a little self-congratulatory about this as I walked back through the windstorm to the cottage. The diary for that night shows two separate entries. The first is in ink.

Jan. 2nd: Terrific winds this evening.

The second is in pencil.

I understated it. An hour or two later we had lost the Tunnel, blown right off its moorings to rest draping the hedge, a complete write-off, and, much worse, almost the entire roof of the shelter. We heard a noise indoors and when I went out with the torch again, Sam had transferred from the end to the centre bay and sat there unconcernedly still among the shattered sheets of corrugated asbestos and fallen timbers. Even the great barn timbers had been tossed from the back of the roof right over the front and on to the ground ten yards away. The only roof still intact is most of the stable roof – because of course the stable doors were closed. The hens were unhurt.

Sam-Ram came out of that awful night quite unscathed, but thereafter his appetite for any food other than grass became almost frightening to see. There was no peace for us. No sooner did we show face outside the cottage but his deep 'baa' carried demandingly across the gate to us. We could not enter Sheepdip to start tidying up after the storm, or to feed the ponies, to feed the hens, to count the sheep, but what he was at us, biffing our thighs insistently with his hard black head, bleating, worrying. I fed him his due ration twice a day but his appetite was unappeased. If I scattered poultry corn across the ground for our dozen Light Sussex, Sam collared that. He always stole half of each pony's food and he nudged us continually for more, more. Never now did he go off to graze with the flock. Something was wrong, and perhaps because it had always been his feet that were wrong, we decided we had better give them more treatment. Perhaps they were so tender that he could not walk far, we thought.

I had him follow me into the stable for nuts, closed the door on him, took hold of him ready to negotiate him into the cradle. And I gasped at his thinness. Even as he nibbled at my fingers for yet more sustenance, my other hand could feel no flesh anywhere under his scruffy old overcoat of a fleece, nothing but bones. He was absolutely nothing but a bag of bones.

I was appalled. Our great old Sam-Ram, how could it have happened? He shamed us, and he even grew worse. His hunger was a madness, yet he would eat no hay, no grass. He started to scour badly, his eyes dulled, his fleece grew ever more staring. Ashamed, alarmed at last and almost too late, we fussed around him daily, bedded him down in stable on clean straw under our temporarily repaired roof and called in the vet. Antibiotics twice a day and complete denial of all food obtained a respite; cutting and carting and his subsequent nibbling of brambles and ivy granted us a little hope. Yet he must have been very close to death's door. His scouring continued. He was so very weak. He ate nothing. Twice I saw him stumble and fall as though from sheer weakness.

'The mark of the tiger' was upon him.

He had eaten nothing now for days. His scours ran from him, staining him. Perhaps he was very old, I would stand by him and think. Perhaps his time had come.

"Worm him," said the vet, next visit.

Sam-Ram did not struggle against the good dose of worm medicine we pushed into his poor old muzzle, but just sat there in dignified acceptance. And it worked. Soon the scourings positively burst from him, but then ceased. He rested. I brought him a little ivy and he nibbled at it quite delicately. His teeth were still quite good, all there and certainly capable of plenty of grazing yet, could we get him to it. He rested some more, surrounded by his Light Sussex companions, until I persuaded him to his poor feet and ushered him gently out. Stiffly he left his hospital-stable and I felt we should be one on each side helping him towards the gate of Sheepdip, which I had opened so as to offer him the complete freedom of my lawn and indeed my entire garden. Overwhelmed by his twenty-yard trip from the stable, or the considerably more exhausting one back from the brink, or perhaps only by my gesture of remorse, Sam-Ram sat down on my lawn and considered things. Soon, as he lay, he began to nibble the fine grass.

Day by day, now, he came out and ate my lawn. Night by night, he returned to the stable. You could feel every knob on his backbone and every rib, sharp and fleshless still, but we had begun to hope. A damned clapped-out old ram, he was, no good to anybody. Those dark brown stains of his past scourings, the lifeless wool of his fleece, and his terrible remaining weakness were a disgrace; they mocked at us, and all these hours we were spending to nurse him. I brought him cut Bramleys and chopped carrots as delicacies. Sometimes he ate, sometimes he just sat and considered. The large cob nuts which once he had been so ravenous for, now he simply could not eat at all. When I offered him the much smaller pony nut, this sometimes he would take into his mouth, but then had no strength to eat. We would sit together, considering ourselves, within sight of death still.

A day came, however, when I knelt before him with a large Bramley apple which I cut into sections and offered him piece by piece at a nice slowness. He ate it with relish. I fetched another Bramley and as I sliced it, he ate that one too. Nine fine Bramleys he ate that day, and soon he seemed to fancy a turn on the lawn. Until now he had done no more walking than the twenty yards from stable to lawn and that only on the days of open weather. This day, however, he walked and plucked the grass for a good hour. His droppings were almost normal at last; it looked as if he had decided to live.

Towards the end of January, one windy afternoon, he decided not to return to the stable for the night. By now, it seemed, we had come through something together; a certain friendship seemed to have been formed. "Baa," he'd say when we met, coming to me well-manneredly and walking at my side, entirely different to the rushing and butting when the worms had been devouring him. So that windy evening, Sam settled himself down just inside the old garden gate, under the tall and close conifers, on a quilt of brown twigs and needles where he was out of the wind and where no sheep had ever been allowed before nor would again, after Sam. Next morning he was still as comfortably ensconced there as he had been when I had locked up at night. He had behaved admirably, ravishing no plant in the garden, and rising from his bed only at my bidding, once again to take a turn on my lawn and begin to eat his way back to strength.

Still he could not find the strength necessary to chew the bulkier concentrates, but I tried him with just a little chicken food in tiny pellet form and this he could eat, although still with some effort.

Gradually he began to accompany me everywhere, all around the house and garden, even round the back to the covered store where we kept the tools and feedbins. Day by day he became more dog than ram, even to the extent of feeding out of the same bowl as Sam-Boxer, to us an unbelievable thing. He grew strong and graduated off of small amounts of poultry pellets on to pony nuts. He outlived January, slept in the sheltered places around the house, respected all my plants and had only one more crisis to outlive.

As a symptom of resurrection, he now began to divest himself of that awful tatty old fleece and to grow a new dark sheepskin coat underneath, and that right when winter was at its most bitter. What at first had been only a slightly torn patch of wool on his shoulder soon became a wholesale shedding of large portions of his entire fleece, a sloughing of lifeless wool under which the dark felt of his new fleece had hardly started to grow. Poor Sam was a scarecrow of a ram. Nevertheless he grew stronger all the time and was being weaned back out of the garden and into the flock for longer periods each day. He had by now of course established territorial rights to my garden as well as this valley-of-the-shadow-of-death relationship with me, and therefore he regularly presented himself twice daily for his half-pound of nuts at gates and various strategic points adjacent to the garden. Should we be tardy in answering his bellowed demands for food, he grew belligerent enough to try to do things the hard way, thus tangling with bramble and wire which further loosened and pulled away that awful hanging fleece.

"You poor old bastard," I used to say affectionately at dusk on the really frosty nights. "Come and shack down in the shelter round the back of the house," and there he would settle down on old paper feedsacks, between my only two bales of straw, whilst I draped his bony back with spare hessian.

At morning, small black beads of his dung necklaced my clean concrete floor, announcing improvement in his general condition, although his total lack of sheep-nervousness also announced that he still had far to go. Without flinch or protest he allowed me to dress his shabby feet with wads of cotton wool soaked in formalin and to enclose this in new polythene, and I really believe he was willing to deliver himself as completely into my hands as a suffering man does into the hands of a surgeon.

His feet hardened well. His thinness was still painful to see but he was eating normally and his undercoat was visibly growing. The old dead wool peeled and fell off, all but a patch on his back, and he long looked the poorest of creatures. Nevertheless, he was alive and coming well; for the time being, we were content.

Sheep were mainstream. Day in, day out, whatever else happened, sheep-counting, sheep-inspecting, sheep-fetching, sheep-treating, sheep-feeding never allowed us a day's freedom. One day's sheep-neglect invariably involved extra work and worry, and only the fact that the garden was sodden continuously from November to March and therefore ungardenable kept me to a daily winter sheep-routine without too much grumble. Little by little Sam-Ram gained enough strength to get back into this flock-mainstream, but always as we went out at morning or prepared to shut up at night, it was Sam-Ram we looked for first. He grazed with a perpetual question mark hanging over his future, of course, but his future was months away, beyond lambing and beyond summertime. He seemed to me to be something of an old soldier; he would never die, but only fade away.

14   A one-eyed peacock

The garden suddenly was running with dogs. As the young springer came dashing up to me with tail-wagging welcome, old Sam-Boxer growled and planted a massive paw across the springer's neck, pinning him to the grass, at which Simon, just climbing over the brow, shouted and started a rescue charge, at which his lurcher and Caleb's retriever joined in the fun while Caramelle, who fortunately was on a lead, grabbed and sunk her teeth well and truly into my ankle and there was a general mayhem of dogs which amused me little, as I explained in brief angry words to my son whilst we calmed them all down.

He answered me with silence. It was normal. His answer left me, as usual, feeling that I really should not have reacted so angrily, that really I attached too much importance to what after all was no more than a quarter-garden and what on earth were dogs for except to run free as the wind. They were *his* damned dogs running free as the wind in *my* garden where I, so help me, even had to keep *my* dogs on a leash in case of incident. Sam-Boxer might have killed the springer and the saluki might have made a real mess of the new lurcher, and so on.

He heard the old fool out with that quiet smile of his, waited within a small silence before saying, "We're just going up into the woods to get a few pigeon."

I nodded and sighed. The two young men strolled off, guns on shoulder while the springer, lurcher and retriever started a new game through my foliage border, screamed round my conifers, bounced on my burgeoning bulbs, leaving me to hang on and yell at my own two excited dogs and to feel that really I had no right to be here at all.

I stood alone and philosophised about the generations whilst Simon or Caleb whistled up the track for their dogs. The reason for living was life. Life demanded further life. If life had any shape at all, it was pointed in the direction of further life, different life. It demanded that child supercede father, that I make way for him. He and his damned dogs were young; I and my damned dogs were aging. In a year or two, all the young men would cross the river with their goats and tents and dogs on their way to new pastures, but I would be too old, unable this season to cross the river. I would sit alone on the bank, I would close my eyes and start to die.

"Simon," I screamed into the wind, "call your blasted brown mongrel off my rhododendrons, now then."

An obedient whistle floated down from behind the house, the lurcher sprinted off across the lawn and leapt the iron gate beautifully as it left and already I was thinking that perhaps it did not matter that Lady had pissed on my expensive rhododendrons, for they really did look a sickly lot. Perhaps I should not worry about that bottom part of the garden. Perhaps I should concentrate rather on the higher, drier parts nearer the cottage. Perhaps I should already, so early in my garden career, plan a phased retreat towards higher ground and then the conservatory, to a tiny heartland where years from now I should potter and mutter and totter over an area two yards square, my utterly last and own domain.

No more than an hour of routine sheep-minding, or seedling-fussing or some such could have passed before those damned dogs and their attendant hunters were suddenly back, once more marauding all over the garden. Dominant male once more, I prepared to beat my chest and have another go at exerting my fading authority when I seemed to recognise that Caleb was carrying something that looked for all the world to my fading eyes – I wore no spectacles that day – like bagpipes. They were happy, Simon and Caleb, I saw, and as I peered at them I realised that Caleb had obviously captured somebody's bagpipes. It was, after all, happening all the time down here. He carried it correctly under his arm, all ready doubtless to break into some devilish lament.

"Where shall I put it?" Caleb asked, smiling still, and as I went towards him, temporising over some suitably ridiculous reply and, focusing my vision more accurately, the bagpipes became a peacock.

"It's a peacock," I said, accusingly.

A peacock, mind you. Here.

They nodded, smiling still.

"Where the devil did you get it from?" I observed the brilliant bird in abject disbelief. All the dogs, too, sat about, tongues lolling, smiling and watching in similar disbelief.

"Up in our woods," Simon said.

How could that be? We do not keep peacocks in our woods. Few people do; no people do. Not in Wales, not in Dyfed, not round here. Wonders of water, yes, castle ruins, yes, peasant farmers and millions of sheep, but not peacocks. There were no peacock sort of houses round here. Ugly grey villages smuggling into wet hills. Geese, now . . . lots of white geese. Perhaps it was really a goose? No; heavens above, it really was a peacock.

"Dogs found it, up in a bush," Caleb said.

I noticed that Simon had a bouquet of peacock feathers.

"Is it all right?" I asked.

"Dogs pulled a few feathers out of its tail, that's all. And it's only got one eye, but it's all right."

The Case of the One-Eyed Peacock.

All right, it *was* a peacock. Caleb's left hand grasped its large legs, his right hand lightly kept the peacock's beak out of his eyes.

"Where shall we put it?" he asked again.

"Better put it in the stable with the hens, I suppose," I answered doubtfully whilst wife, daughter and grand-daughter came out to join in our wonder and praise.

That night, our stable held one dozen old Light Sussex hens, their magnificent rooster, Sam-Sussex, the recovering Sam-Ram and our new peacock whom Phyl even attempted to call Sam also.

Peacock was always a bit too royal for the hens and a great deal of agitated clucking continued too long for our peace of mind, so that we had to transfer him to the shed next to the cottage. There had been no warfare between chickens and peacock, but we deemed it wiser, if we were to safeguard our ever more precious egg-supply, to move the peacock, and so he installed himself imperiously on my old kitchen-table-work-bench, ate nothing, and gazed out at me gazing in with my look of continued improbability. Our wood is such a scruffy wood, our country such unpeacock country; how on earth did he arrive there?

"No, sir, no peacocks have been reported missing, but I'll make a note of it," the police station said impassively.

The duck-man, the vet, Thomas, nobody it seemed knew of any peacocks around here. Peacock must stay. I was content. Perhaps I was the only man in the whole mad world spending his hours tending a one-eyed peacock and nursing an impoverished ram.

I opened the shed door on the second day and crept away. When I came back later Sam-Peacock – it was ridiculous; you cannot call a peacock Sam – was sitting nearby on a field-gate and I started a campaign of feeding and taming him forthwith. For a few nights, fearful of foxes, I was brave enough to catch him up and re-shed him, but it was a somewhat fearsome business to be within range of those beating wings, and after a few exhausting contests I suggested he look after himself at night. Forthwith he flew into a high conifer and roosted there quite regularly. By day he learnt quite soon to come for the corn I spread for him, and then to come to the small tissicking noises with which I accompanied the corn and day by day he settled in, truly proud, truly beautiful in a quite unbelievable sort of way, but so utterly vulnerable in this wild country.

Peacock's brilliant career at Dolgwili was threatened only once, and that by Simon's springer and lurcher who, once more off their leash, came upon the peacock too suddenly, too alarmingly. The great bird took off, flew right off Dolgwili, across the river and the road to land clumsily in the boughs of a sycamore growing from a rocky cliff. The boughs actually overhung the road; the peacock spent half a day watching the cars and lorries whizz past below. Simon, after another round in our particular generation game in which he absorbed all the tongue punishment I could throw at him for being so careless with his dogs, spent half a day also trying to frighten or lure the great bird out of the sycamore. When he had spent long enough trying such unpromising tactics, I took corn in hand and lured our great Light Sussex rooster and his dozen hens down on to the lower slopes of Garden Paddock where we were in full view of the escaped peacock. As I spilled corn and made my little tissicking noises, the peacock grew more and more restless. I threw extra corn about the grass to keep the white Light Sussex there as decoys, and then I retired to have a word with Sam-Ram or whatever. When I returned in half-an-hour, the peacock was back on Dolgwili.

He graced our spring marvellously well. No day did I look on that extravagant bird but that I wondered at such beauty. But we grew cocky, peacocky. He deserted his tall conifer near the old orchard and took instead to roosting in a scruffy hawthorn just above the Spout, a perch just as high in one elevation but much less high in another.

He did not come to my tissicking breakfast call that morning in April. Even as I went to search, my spirit was heavy with foreboding. Through the gate, on the track leading up to his perch, I found the few first feathers. More and larger feathers strewed the track as I trudged on. It grew worse. I bent and picked up one small exquisite feather with that green-blue eye of fantasy dyed on it. In the mud near the spring of the Spout were half-a-dozen of his long tail feathers. Feathers traced half-a-circle round the slope of the meadow beyond. His heart-breaking carcase, fox-savaged, was deep in the hedge where in terror he had tried to hide himself from the snapping jaws.

I wept. I turned away without heart to bury my dead.

"Is there anything you want done?" Simon innocently asked that evening when he came to Dolgwili to exercise and train his dogs.

"Only one thing. Find and shoot the fox which destroyed the peacock last night."

"No!" He gasped with more shock in his voice than I had ever heard him utter before. He stared at me desperately and we looked at each other as if we had not always known that one day a fox would kill the peacock.

We wandered about, desultorily picking up the gorgeous feathers and examining them as though those colours could possibly tell us anything about anything. Peacock should never have been at Dolgwili at all. This is a rough, fifth-rate place, fit for sheep and foxes but not for peacocks.

Soon Simon wandered off with his dogs and gun into the woods and as I watched him go I remembered the days, not that long ago it seemed, when he quite worshipped the fox. In those days he loathed hunters.

He found no fox. But that night it returned and carted off the remains that I had not buried, leaving more flurries of feathers all across the grass, which all of us collected. To this day a large glass jar full of peacock's long, incredible feathers, still stands, gold and blue and green, delicately moving in the draughts that blow through the shelter at the back of the cottage.

15 Why do you keep sheep?

"I'm sorry," I told the man fresh from the city, "that I blew my top so unpleasantly in front of your young family just now."

"That's all right," he smiled. "I saw what happened. They know the word. I understand."

So I stood with him, basking in the early evening sunshine on the terrace. It was almost a ritual for visitors to stand like this. One could almost hear city folk unwinding, during those first few days of holiday as they slowed down enough to start seeing their country surroundings.

"Our son brings his dogs here most evenings for exercise and he will *not* put them on a lead, so they come rushing and jumping all over the garden and of course this evening it was exactly at the wrong moment."

"Yes."

I had just been ushering Sam-Ram into the stable – for routine worming this time – when Simon's brown lurcher had leapt with abandon clean over the field gate from the garden and panicked Sam utterly. The ram had rushed about blindly and finished up determinedly butting a wire fence, quite berserk and out of my patiently-built control. Completely involved with the well-being of the ram, I had shouted commands most forcibly over my shoulder towards Simon without realising that the city family was just arriving home for dinner.

No great harm had resulted. Simon had turned immediately and slunk away with Lady not so much angrily as disapprovingly. Being a calm and silent man himself, he disapproved of my own lack of calm and silence in those agitated moments of mine when anxiety came bubbling up through my head. Long long ago he had reported critically to his mother that Dad was useless to go out walking with because he would talk all the time and keep calling the dogs. What secret life can a son discover and enter along the hedgerows or the river with a damfool father chattering away?

Months ago, close to death's door, Sam-Ram had been known more than once to share a bowl of food with a boxer, Sam-Boxer, who himself, back in Suffolk, was something of a sheep-chaser. Now, arrived on the return journey at life's door, all Sam-Ram's true nervousness had returned and he had turned and fled from the bounding lurcher as a healthy sheep should – quite unnecessarily, for Simon had long been training the lurcher to be trustworthy among sheep. Sam-Ram was the same with me, too. Where once he would allow me to touch any part of him, to manipulate those poor old feet into those comic polythene bags or to rub a sympathetic hand across his bony skull, now he was quick to keep the correct sheep-distance from my touch.

"Why do you keep sheep?" the city man asked, and I saw that it was not really a personal question, that it contained no semi-hidden suggestion that I must be extremely foolish to waste my life this way. I smiled, seeing the gist of his thinking, and with some sympathy, for it was not that number of years ago that I too had finally sorted through my muddled thoughts to get to the same question. Why, he meant, did anyone ever keep sheep, for what reason?

"Because," I now answered him, "you eat their lambs."

It would have been kinder had I said, "So that I can eat their lambs," but perhaps it was the wish to illuminate that made me use the accusatory 'you'. Nevertheless, he smiled in acknowledgment that the point was taken.

Across our summer pastures in this happy time of year, the season's crop of lambs frisked and gambolled without a care in the world other than the occasional vital one of finding their mother. Now they formed themselves into juvenile gangs, chasing and bucking in sudden madness before a sudden prick of thirst turned them back to mother and ferocious sucking. Soon they would all be dead. As yet the summer need contain no threat of such direness but come July's end we would have to face up to the need for the marketplace. I had no stomach for the slaughterhouse, nor even eyes. I would turn and face in the opposite direction, would even consider the signpost to vegetarianism, as long as I could. Yet all roads lead to death. One has to accept; there is no other way.

Recently, rheumatism in a war-damaged shoulder had directed me to a health-food shop for some herbal tablets, recommended by similarly afflicted friends. I was, anyway, temporarily attracted to unorthodox cures, as indeed to vegetarianism and muck-mysticism generally. Inside the shop a vegetarianism tract had been stuffed into my hand, together with the tablets. It contained over-frenetic condemnation of farmers who sinned as much by producing meat as did all the unenlightened by eating meat. Having read the extravagant condemnation, and feeling personally condemned thereby, I was moved to appeal to the most pleasant gentleman behind the counter what then I must do to become righteous.

"What would you have me do with the sheep we own, then?"

"Why, just allow them to live on the hillside as God intended they should." He was a nice man. His face shone with the sunshine of vegetarian salvation.

"The lambs, too? Not kill them?"

"Of course not. It's revolting."

"I agree." I pondered the revolutionary consequences.

"But," I said in due course, "the old ram would serve his daughters, the young rams will serve their sisters and mothers, joy would be unconfined. The hills will be covered with more and more and worse and worse sheep, farmers will not trouble to manage them, will not hedge or fence or feed or dose, the sheep will come into your garden and eat all your precious compost-grown vegetables and . . . "

Really, my mind boggled. There were even people – they seemed to be young people always – who, according to the tract, would not only refuse meat but also milk and eggs.

"Wool," the shopkeeper said. "You could keep them for their wool."

Well, yes, there was wool. "Last year a fleece brought me in about ten bob. Let us assume with the higher stocking rate which your vegetarian regime will usher in, and always remembering that a sheep's greatest enemy is the next sheep, let us assume we shall have ten sheep to the acre and let us assume that the price of wool will double, although it is more likely to halve. An acre will then return, say, ten pounds. Do you think anybody will keep sheep for that? No. I think you suggest an impossibility. The hills are there for ever more, and so are sheep. Nothing else will thrive up there."

He was a nice man in a nice shop. It was coffee-aromaed, and filled with good food and lovely ideas, and after words had failed me in pursuance of the vegan nonsense, I confessed to such splendid practices as enjoying muesli and yoghourt, of composting all our kitchen waste with hossmuck and nettles, and I recalled that no apples in all my wasted life, nor almost any other food, had tasted so inexpressibly delicious as the Cox Orange Pippin I had grown in my first garden with generous mulches of donkey-and-duckling muck with lawn-mowings. I had further, I mentioned, withdrawn my interest in Dr. Rudolf Steiner's teaching only when regrettably I had been unable to obtain a hart's horn in which to bury my starlight or whatever, but that I was persisting with wearing a copper bracelet against the rheumatism and my mind was still sufficiently open to wonder at least to consider *any* unorthodoxy. But still, if he and his friends ever succeeded in persuading the world not to eat, from July on, the splendid creature the world adores in March, then certainly on to the list of endangered species would go one more – the sheep.

All this I related to the young city man in the quiet and anticipatory hour before dinner as we strolled across Sheepdip and among our flock. I recognised in him something of my younger self, trapped by the needs of his adorable young children and wife, and also by the demands of his own competitive nature to rise ever higher in the competitive hierarchy of his tribe, yet at the same time hearing ever clearer and closer the siren voices of our Welsh hills and an old way of life. This one wanted, just as we ourselves once had done, to breed dogs, red setters, to buy just a few acres down here, to have kennels and thus to escape the turmoil of cities. This man, or one almost exactly similar in circumstances and dream, came to us regularly through the summer weeks to refill our leaking money-bucket by paying, as we thought, handsomely to sleep in our modest bedrooms, to talk long into the evening and to sit dreaming about his forthcoming escape. They never escaped. They read the local newspaper, they looked at properties for sale, and at the end of the week, sighing over the heads of their children, they packed their expensive bags and went back to their expensive suits and their claustrophobic cells.

"So how many sheep have you got?" he wanted to know.

I pointed out Sam-Ram, and his oncoming son, Samson, and named each ewe for him, a little inaccurately no doubt, for most of them had been named by Phyl and to her the daintiness of Dainty, the gentleness of Gentle and the nice fleece of Nice Fleece was always more apparent than to me.

"So that's thirty-four altogether, but nineteen of them are this year's lambs and soon they'll have to go."

Yes, it had been quite a good season, I agreed; nineteen lambs from thirteen ewes was fair enough. One very large lamb had been stillborn and one of Simon's lambs, out of his half-uddered Sad One, had died, but overall, yes, it was satisfactory. Nevertheless, our sheep-farming was not as happy as we would have liked, and we had drifted into a time of reconsideration. Over mid-morning coffee and in other domestic lulls, our thoughts continually ground towards sheep-change. Of the nineteen lambs delicately nibbling these young grass tips, twelve were ram lambs. Their sex alone determined their fate; no pale philosophy tenderly dreamed up by my vegetarian shopkeeper's friends could save them. The simple biologic that only one ram is needed for forty females was a brick wall against which many heads had been and would for ever be knocked. We had ourselves arrived at that same brick wall, and although we did not knock our heads against it, we did not like its face one little bit, for plainly writ across the wall's face was the sign: Slaughterhouse. There was nothing very attractive to us in the thought that we were breeding lambs for slaughter.

Not this week's city man, but perhaps last week's, had been, on our recommendation, to a nearby beauty spot to see the splendid waterfalls and rapids on a lovely small river by a very old stone bridge, the place famous for its fish as well as its beauty. Frequently salmon and sewin could be seen leaping the falls on their way upstream and here too the coracle was still to be seen on the river. Fishing by day-ticket even was allowed here, so it was such a delectable place for men, women and children that it was steadily being destroyed by those same men, women and children.

The lovely grey stone slabs of the river near the bridge that day were lined with camped-out anglers. Every few yards a man sat casting his worm or whatever into the black-blue, white-laced whirl of water. Grey wagtails landed on river rocks and danced a bit. Children piled out of cars to stand gazing at the power of the cascade through the rocky gorge, and tinny transistor radios blared forth to make slightly more tolerable the intolerable natural beauty all around. Grey-haired matrons posed with grandchildren for instant photos, cars arrived, cars departed, and amid all the Hogarth busy-ness of the scene, it became noted that one rod of the many constantly moving rods was almightily bent under the weight of a fish. Commotion. People stood up from their picnics on the rocks, photographic poses broke, children shouted and moved nearer. Our city man joined the forming crowd to watch the fun, to see the angler catch his salmon.

It was a fine fish, he reported to me, a large and beautiful fish. And the angler, when he had landed and unhooked the splendid creature, took it by its slippery tail and however else he could strugglingly grasp it, and he began to bash its head on the grey rocks. It was not easy, it was not pretty. He swung it as best he could, awkwardly, repeatedly. He bashed it until it was ugly in its death. The children and the sensitive in the crowd turned away in distress. My city man and his family climbed silently back into their car and drove away.

The death of an animal, no less than the death of a man, demands respect and, where possible, some mark of that respect. Particularly does it demand your respect when its death has been brought by your hand. Countrymen, and shooting men in particular, acknowledge this. Similarly did I require respect, and some mark of respect, for our lambs, when their turn came. Well, it was a foolish notion, of course. The market has no time for respect, and its dealers no need. I came in time to consider that the Oriental approach, that of ritual, of ceremony, of celebration, was far more gracious than ours, and it may be that the marketing of lambs one day will be marked by some quiet ceremony of respect beneath the conifers at Dolgwili.

In all fairness to the marketplace, it is impossible not to be grateful to the men who have to spend their hours there. Somebody has to buy surplus lambs, somebody has to kill them.

There is no escape from that. However we argued against it, always we came face to face with our own squeamishness, our own rather precious over-sensitivity in the service of death. Simon, on the other hand, quite properly dismissed our squeamishness and could face up to all the requirements, killing and dressing any creature with calmness but always, I begged him, with due respect. If we were willing to eat lamb, we should be willing to kill lamb, Simon argued, and I could not but agree that we should. But we weren't.

"That one, the blackfaced young ram there, that's Samson, the one we hope to keep as successor to Sam as our stock ram," I had pointed out to the city man. Samson represented our first tentative and as yet somewhat unpromising step towards the only alternative to rearing lambs for the meat trade, that of rearing lambs for breeding. It was still only a partial alternative, for all sheep-farming eventually rests on the single fact that we eat lamb. Nevertheless we were discovering by visits to the market that a good ram lamb, of desirable size and breeding, could fetch sixty pounds, whereas a fat lamb for slaughter would fetch only thirteen or fifteen pounds. Samson was a year old and as yet quite without his father's dignity, but equally without his father's terrible feet. Nevertheless Samson stood there as a mark of our growing confidence in ourselves.

Our former attitude of letting the Dolgwili sheep get on with it, come what may, had almost gone. Sam-Ram absolutely must go, we told each other again; the ewes absolutely must be inspected and, where necessary replaced, and maybe we should consider inoculating, castrating, docking and even dipping, and all those other operations we had previously avoided through ignorance or unawareness or timidity.

"But," the young city man said doubtfully as we turned for home, where Phyl was sweating it out over her hot stove, "is all this a financial proposition?" He was in banking, you see.

"Not really," I laughed, shaking my head and closing down defensively. He would be asking why we did it next. "No, but we enjoy it."

I opened the stable door and Sam wandered out with a deep 'baa' of disgust that I should not be offering him something special to eat. He began to graze. His new fleece was growing quite nicely.

"What about wool? Do you do your own shearing?"

It was tempting to reply "Not yet," but in honesty I had to say, "Oh no. Two young lads from the village come round. They're supposed to be here this week. They've got a little petrol engine which they cart from farm to farm. I must phone them up again."

As the banking man and I walked back to the house, I could hear his mind calculating, and even the questions he was not asking. I could even hear my own silent answers, and I thought how absolutely watertight and safe his own way of earning a livelihood was, and how ridiculous it made sheep-farming seem.

16   Gid, Orf, Fluke, Braxy, Strike

Disease, like death and divorce, was something that happened to others, not to us. Persuaded still by the original optimism of the ministry man, and recalling our first trouble-free years, and further assuming that some private god continually guarded Dolgwili against evil sheep-spirits as of right, we had come to believe that our sheep must be immune to that immense list of awful things that evidently happened to the sheep of others. Gid, twin-lamb disease, joint-ill, orf, fluke, ticks, sturdy, scrapie, pine, swayback, braxy, daft-lamb disease, black disease, blackleg, louping ill, enterotoxaemia, and at least another dozen other ills were not for us; did not our sheep live a simple and natural existence on sheltered hills in the cleanest air, straight off the Atlantic?

"Surely, living naturally like this," Phyl had said a dozen times, smugly persistent in her belief in our rightful immunity.

Tiny pimples on the black muzzles of our lambs had suddenly spread and joined to become festering sores almost before we had become aware of it, so complacent had we become. It was orf, the vet confirmed when at last I rushed a typical sufferer down to him. It could spread all through the flock, he warned. Lambs could pass it to the udders of the ewes; it could be a most unpleasant business. Fortunately it yielded to spray treatment quickly enough, but treatment involved almost continuous inspection for new cases, and to inspect at all, first catch your lamb. Once again, the crazy business of trying to run sheep without the help of a sheepdog loomed daily for us as first one rheumaticky pseudo-shepherd and then another pretended to an agility long since lost, as we slithered and grabbed and fell about in laughable efforts to catch our frisky lambs. We bought ourselves a crook, we conjured fiercely cunning schemes of catchment, we bickered gently, but day by day the orf got away from us and spread nastily, simply because we had never yet perfected a system of catching up and holding our sheep.

Our young men's methods, when Sunday by Sunday they had been attempting to banish foot-rot from the flock, had been to stroll with irritating nonchalance to the outside of the flock and then little by little drive them down towards the three-bay shelter where the sheep were to be held. Their drive had been characterised by this continuous nonchalance. Without a briefing or plan, they wandered across the pasture to stand out there, no doubt yawning slightly and examining the skies for woodpigeon or woodcock or the slopes for hare, before turning and almost accidentally, it seemed, slowly moving the sheep whilst Phyl and I waited in the wings clucking like old hen and old rooster about their wayward youngsters and wondering what sort of vanity possessed them to display such total unconcern over so vital a job. Even when the sheep broke, split and galloped away so that the gathering had to start all over again, Simon-Caleb managed to maintain total casualness, and any elderly suggestion that a wire fence should be erected temporarily to funnel and trap the sheep was rejected with even more unconcern. No doubt our respective motivations during that period were quite different could we only have known. We were there only to ensure the banishment of foot-rot, but they were more probably there to ensure the perfection of their gundogs. Labrador, lurcher, spaniel and terrier were never far away, all the time being trained to stay there, to heel, to sit, to what-the-hell-do-you-think-you-are-doing, Toff, and above all never to worry the sheep, for without that they could never be allowed to go hunting.

Well, Mum and Dad now could fuss away to their hearts' content as they planned and slowly built their fine solid paddock of trustworthy, upright posts sunk thirty inches into the ground and unassailable cleft-chestnut fencing. It was all square and admirable, worth every last ache and pain that it caused in every muscle of our two sore bodies. Just as the kitchen stove was the heart of the home, the new paddock was to be the heart of the farm.

Farmdogless still, shepherd and shepherdess began to perfect their technique of leading and driving their sheep. Phyl,

rustling a polythene bag of sheep nuts and calling, "Tot, tot, tot
. . . " would gather them about her and slowly move off towards
the gateway of the paddock whilst I, most professionally clad in
shepherd's crook, rounded up stray lambs and followed until all
were safely paddocked. Lamb by lamb then we could catch
them up, inspect their sores and spray the gentian-violet re-
deemer across their noses and our own clothes until little by
little orf was contained. Or almost . . .

A day of crass foolishness arrived in my life when in one
morning I pruned roses, cut and filed the donkey's hoof, and
treated lambs for orf. Why my shrub roses have thorns more
wicked than the tame hybrid-tea is as unanswerable as why I
worked among them without gloves, but that is the way it was
that morning. The thorns scratched and plucked the backs of
my hands quite drastically, but I heal well and I drifted from
job to job just as thoughtlessly immune as we had always
supposed our sheep to be. I drifted to the absurdly-named
donkey, Marigold, whose feet were beginning to suggest that
she had taken to wearing winkle-pickers and which I began to
suspect might be a little unhealthy too. From there a further
drift with my lady took us into the orf-zone. In all that time I
had not even washed my hands.

"I sit here," I confessed to the doctor in the village next
morning, "and marvel at my utter stupidity." I related to him
the details of that utter stupidity, showing him my swollen and
throbbing hand. The swelling was spreading fairly rapidly to
my wrist.

"Could it be orf?" I asked, smiling to show I was hardly
serious.

"Oh yes," he answered as he wrote the prescription for one
more batch of the red-and-black capsules which had become, it
seemed, the universal cure-all.

Hands and wrists ignored the cure-all, continuing to swell.
Fingers grew sausage-like, and large sausages at that. By next
morning they were so bloated that I could do nothing but hold
them uselessly up on my chest somewhere. Worse even than the
useless sausage-fingers was the stigma in the mind. I had orf, a
sheep disease, a most unpleasant sheep-disease too. I began to
think I smelled like a sheep. I began to suspect my feet. Un-
dressing fingerless in the bathroom I found myself surrepti-

tiously peeping at my toenails for any resemblance to Sam-Ram's, and then even to sniffing around a bit down there too. I sat to ponder on joint-ill, strike, braxy, sturdy, scrapie, to wonder just what the sheep had in for me next. When I looked in my mirror, I really had got small red pimples beginning to show around my mouth. They were facts as certain as my horrible fingers, but fantasies crawled in the mind. I was ill with sheep.

Well, the good doctor consulted with his antibiotics in some alarm at the spread of my orf, and to good effect, for under new treatment it subsided satisfactorily, but from then on a chastened me realised full well that none of us is immune to anything. It was time, past time, that I knew this. Circumstance rules; we are victims, all of us. The accident of the night has already decided what we are, the accident of the day decides how we go.

The new ruling circumstance was heat. The summer had grown hot. Ewes and lambs spent their days in shade. They liked the cold cement floor of the shelter and the daylong shade behind the shelter. A few of the lambs took to misery, tucking their noses down into the grass, not moving, not grazing. Occasionally they would glance backwards and turn as though towards some irritation tailward. A darkish stain showed on the fleece. One had a slightly torn fleece and many of them had black soilings round their tail and behind from scouring on the new grass.

"Kouh! Just come and look at this," my fastidious wife commanded me, her voice arched in disgust. Together we held and examined one of the most miserable lambs – with gloved hands now. The fleece was absolutely crawling with maggots. Parts of the fleece, dark at the roots, were already destroyed and coming away. As Phyl's fingers parted the wool and searched, more and more maggots were revealed, all crawling and burrowing and feeding like gentles in an angler's tin.

Thus and thus was our immunity mocked by orf and strike. Both are easily controlled, both are seasonable, but we had never seen either before. I began to presume that the good fortune of our first seasons was simply due to the fact that Dolgwili had enjoyed some years of sheep-rest during which it had rid itself by rain and frost and sun and time of those

129

parasites and pests which prey on sheep. Now with the return of sheep and their retention year after year, disease was brewing up again.

Perhaps those first lambs were lucky. They had suffered worst under the strike of the blowfly and thereby now had lost considerable patches of fleece, but as the summer grew ever hotter, perhaps they remained cooler than their mothers who panted and lay prostrate, waiting and waiting, with us, for the shearers to arrive.

"Who is it speakin', then?" the woman would lilt. "Dolgwili? Ooh, yes, all right. Listen, the boys are out, see, but now then, the moment they get back I'll tell them, see? Aye, aye. Now then, yes, I've got it, Dolgwili. Aye." Bonk.

She never said goodbye. Bonk, always the phone went.

"Who? Dolgwili? Oh, aye, aye. No, not now. Be back soon, they will, aye. Now then, try again, will you? About eight. Aye." Bonk.

"Ooh, yes. Yes. I did tell them, you see. Aye. We've got a note somewhere, you see. What's your number again? Good. There we are then. Sometime next week. Aye. Guaranteed, myn." Bonk.

"Ah yes, now. I did, didn't I? Well now . . . Busy we've been, see. Well, now then . . . We'd better fit you in. How many you got? Aye. Would Monday suit you? Aye. May be late. Aye." Bonk.

Of course it would be late. Last year they did not arrive until nine-thirty and did not finish till the moon was up.

"Christ, yes, I did, didn't I? I remember now. Aye. When's your collecting day? Next Thursday? Oh, oh, there we are then, plenty of time. No, no. No, we won't let you down. Good boy. How's Simon, then? Got any puppies for sale, has he? No, no. Aye, well, I'll give you a ring before we come." Bonk.

"Who? Aye. Oh Dolgwili. No. Both out. Aye. I'll tell them."

Day by day, the sheep grew more uncomfortable. The summer crept towards collecting day inexorably and every damned time we phoned, yes, they were coming tomorrow, or Monday, or next week. The whole wool business which had started with such promise of efficiency was, as far as Dolgwili was concerned, steadily going bankrupt. A card had arrived notifying us that our wool was to be collected from the Rock and Fountain on Thursday July the Somethingth, that a wool sack and labels had already been sent to the pub for us. The wool would be weighed, loaded on to lorries, transported for grading and eventual selling, after which we should receive a cheque from the Wool Marketing Board.

"Did you remember to call in at the pub for our sack, Simon?"

"None left. All been collected."

"What again? Damn."

"I'll see if I can get one from the boys."

"Good."

But what is the good of a woolsack if the wool is still on the back of the wretched sheep?

"Ah, hello, Mr. Jones, how are you? Good. No, no. She's very well, thank you. And how is your mother? And your uncle? Still got Juno, has he? Good. No, no, it's me that owes you money, I'm sure. Anyway, the reason I'm ringing is this. Have you seen the two shearers from the village? No. No, they keep promising but never arrive. You've had yours done? Good. No, I've only got days left before the collection. Next Thursday. Yes. Tell me, what happens if I can't get them sheared? Do they? Oh dear. Dear, oh dear. That's bad. No, no. I'm sorry to trouble you. Not at all. Yes, I'll do that. Straightaway. Goodbye, Mr. Jones. Thank you."

"Hello? Who? Dolgwili? Oh aye. No. They're out somewhere. Aye. I'll get them to phone you first thing tomorrow." Bonk.

"Hello? Have they not? Yes, yes. Well, I'll tell you, sometimes I run across them in the mart, you see. Yes, yes. Only days left now, isn't it? Yes, yes. Yes, sir. Yes, yes. Yes, yes. There we are then. Goodbye, sir."

Day after day, no shearers. We gave up. Collection day arrived. Ben came running into the kitchen. "Tractor, Poppa. Mr. Thomas is coming down the track on his big red tractor. International." At his age I used to know all the cricketers; Ben knew all the cars, all the tractors.

"Come on then, boy. Let you and I go open the gate for Mr. Thomas' International, shall we?"

The huge red tractor crept carefully down the hillside track towards us. It drew a long trailer loaded with vast fat sacks of wool. All morning long down the road towards the Rock and

Fountain had been passing loads like this, tractor-trailers, Land-Rovers, little vans, large lorries, cattle-trucks. Everybody was going to the ball except Cinderella-Dolgwili. We held the field gate open, and the tractor chugged slowly through and came to a halt.

"Mornin', Mr. Thomas," I shouted against the putt-putt of his diesel whilst Ben, his hand in mine, stared upwards with intense respect and awe at the blue-smoking monster.

"Very-well-thank-you," said Mr. Thomas automatically.

"How is your mother getting on?"

"Very good, thank you." He had a charming, sudden smile, breaking like sunshine from a face clouded by language-anxiety. "New foal?"

"Yes. A real beauty. Filly. Out of the dun mare."

"Very good."

Hiatus. What else is there clear and simple to say?

"Very warm again. Finished the hay?"

"Aye. Very good . . . Grandson?"

"Yes. Ben. His name is Ben. Say hello to Mr. Thomas, Ben. Isn't that a mighty tractor, Ben? You'd like to drive that, wouldn't you?"

"International." With me it was cricketers.

"Very good." Mr. Thomas revved his engine, indicating the end of the audience. That smile again.

Phyl, suddenly here, smiled a better one back and shamelessly shouted, "I suppose you don't know anyone who could shear our sheep for us, Mr. Thomas? Our shearers have let us down."

"Not been?"

"No. We keep phoning and they keep promising to come, but they still haven't." Only now did she remember to enunciate her words more carefully.

"Duw, duw, duw, duw," Mr. Thomas looked most shocked at such untrustworthiness and his head shook from side to side in heavy disapproval. He turned and looked over our hedge to where our sheep rested in the shade of the old orchard trees, and smiling and nodding in farewell, he took the busy tractor away down the track whilst we closed the gate and sighed a bit. We stood and watched him and his load of golden fleeces down on to the road, soon to queue up with all the other loads, to drink a bit and gossip and laugh in the sun whilst we were left brooding on the crassness of sheep-shearers.

A couple of hours later, the boxer and the saluki started raving in the kitchen. When I investigated, Mr. Thomas and his unladen tractor stood there, ill-at-ease. He would come back and shear our sheep himself, if we wished, after dinner.

"Oh but really, we cannot trouble you, Mr. Thomas. You have so much work of your own to do."

"It's all right," he assured me, most quietly, beautifully.

"Well, we really would be most grateful, of course. That would be marvellous."

"Very good."

"Thank you."

He climbed aboard the tractor, I held the gate open and he chugged away back to his farm. No man ever gave me such an impression of capability. Two hours later he returned with his nephew and all his shearing gear. Our sheep waited ready in our new paddock, and nearby Mr. Thomas spread his tarpaulin, oiled his shearing head, started his little motor and signalled to his nephew to bring the first ewe. We were saved.

They worked hard for three hours in a broiling sun, pausing for beer only occasionally and showing distinct unease when Phyl and Trudi came to watch their exertions for a while, yet methodically and conscientiously slogging the job out. The only labour required of us – and indeed the only labour we were capable of in the processes of shearing – was the catching up of each ewe in turn ready to pass it to the young man, who had a rather pretty dancing technique of getting it across the tarpaulin to Mr. Thomas. The nephew accepted the ewe by taking one front foot in each hand, rather in the style of old-tyme dancing, holding it almost upright and gently urging it along with the help of his knee, a sort of pastoral minuet quite graceful and entirely effective. Had I have been foolish enough to comment thus at the time to either Mr. Thomas or nephew, much unease and misunderstanding would have ensued, and indeed an uneasy beginning to this session had already reminded me to keep my loose mouth tighter.

"It's pretty difficult to make a fleece neat without a tail to tuck in, isn't it?" I had conversationally opened as I watched the nephew rolling the first fleece, which was a tatty one anyway.

133

"Haven't I done it well enough?" Emlyn had immediately asked with plain alarm showing across his young features. As if *I* knew anything about rolling fleeces! Hastily I had reassured him that I simply did not know what I was talking about, and thereafter struggled with my sheep in silence, aware of how different, how much more capable and excellent these men were at sheep than myself.

This great good-neighbourliness that Mr. Thomas was bestowing on us was traditional and it grieved me that we could think of no way of returning his help in kind. Years ago, before such as ourselves had brought our ineptitude and Englishness to this scene, each farm had a close and formalised inter-dependence one upon the other for labour and help. The use of bull or stallion from the larger farm would be repaid by, for example, a set length of potato-planting, whilst seasonal chores like haymaking, shearing, dipping could only be completed on any farm with the help of neighbours in turn. A recognised form of job for job, help for help, meant that money did not change hands in the old days and that the whole community must have been close-knit and in many ways most admirable. I, however, with few country skills and no Welsh and little real experience, could do no better than to repay our neighbour now with cash. As they packed all their tools and folded away the tarpaulin ready to hurry back to their milking, Mr. Thomas and nephew Emlyn accepted most graciously what I most deferentially and gratefully offered them and paused only to consider our lambs before they went.

"Small," Mr. Thomas pronounced them after some fingering.

I made a grimace of disappointment. "Not ready for market?"

He looked about him again, fingered one or two again, and shook his head in a mystified way which caused me to tremble inwardly that I had been foolishly guilty of some elementary aspect of flock management.

"Small," he said again, although when I asked him were any ready for market, he did indicate three which were. I marked them, thanked him once again for his generous good neighbourliness and allowed him to depart. As I watched him and the trailer climb up the track towards his own farm, I was all too conscious of the great differences between us, even though we both kept sheep on roughly the same area of hillsides.

Evenings later, the day's work done, the phone rang.

"Dolgwili? Just coming. Okay? Be there in twenty minutes."

"No, no," I had hurriedly to shout. "You're too late. Mr. Thomas did them for me."

"Oh, there we are then. Good." Bonk.

17   To market, to market, to sell an old ewe

I had already completed the preliminary reconnaissance of my enemy, the marketplace, mixing with the laughing Welsh farmers and white-smocked butchers, listening to the auction-eer and idling around the pens in glorious anonymity, and thus with no chance of making a complete sheep-fool of myself, even when I became bold enough to ask the occasional question of a market official. The chatter was almost entirely in Welsh, but the bidding was in rapid Welsh-English which very soon be-came fairly intelligible. The mechanics of the marketing process seemed straightforward enough – where to enter, the documen-tation, the queue, the weighing and grading – and in any case I had observed that so long as I admitted my ignorance then I was treated with the usual delightful Welsh desire to help and even with sympathy from market people. On D-day minus one, therefore, I was prepared for the marketplace ordeal, and, in my new act of going it solo, I was slightly put out by Simon suddenly buttonholing me to say "Old Dai wants a lift down to the mart tomorrow. He'll show you where to go."

He spoke a little persuasively, even appealingly; he wanted me to take Dai. I saw that all his unprofitable dealings with dear old Dai had done little to loosen the ties that held him to the old rascal.

"All right," I agreed with such lack of enthusiasm that it immediately gained me the small profit of physical help.

"I'll bring him down from the village and I'll load the car up for you."

Into our heroic little Renault next morning at nine, he loaded the three fat ram lambs, Scratcher and Disappointment, the two old and failed ewes, and finally, most difficult of all, the old man himself. He addressed me as 'Your Worship' repeatedly as

136

I drove carefully down towards the town, and we carried on conversation at that same level, dry, totally insincere, slightly farcical. His bony hands clasped the thumbstick with which he had been born, and he sat bent like a trapped scarecrow beside me. I felt sure he knew to the nearest 'thou' how many beans made five and exactly how to sell any that were bad – or had only half an udder. From my reconnaissance I knew the mechanics of the mart, but of the inner workings, of the mysteries, of the warfare of the mart I knew that I knew nothing. Dai did. For that, if for little else, I duly honoured Dai.

Dai was really Simon's man, just as Glyn also was Simon's man. As I lightly talked to the old man this day, I sensed that he was to sheep what Glyn was to fish. Both men had become adept at living within a private crevice, there to obtain a thin living, as might some specialised form of seashore life. They each had, it seemed to me, intimate and hard-earned knowledge based on a huge respect for the creatures to which both had devoted their lives. Both lived in our village, both were so essentially Welsh, both were Simon's men. He would have given both of them his total respect, I felt sure. Whereas Dai's sheep exploits that summer evening seemingly so long ago, with his grand-daughter, had, binocularly viewed, been a direct factor in tempting me into this crazy sheep-farming, Glyn's river exploits were almost completely unknown to me.

Long ago Simon had come home one day with some story of meeting a tipsy man down by the river, who asked him to be lookout whilst he took a fish. This tipsy fellow then had precariously entered the river, wading in ordinarily clothed, slowly to head upstream while Simon followed a parallel but more comfortable course along the riverbank. This man, Glyn, knew all the rocks, all the overhung banks and tree-roots along our river, every traditional hidey-hole of the salmon and the sewin. With Simon as alert and glorying witness, and possibly even apprentice, Glyn worked his way tipsily along the shallow river, every traditional hidy-hole of the salmon and the sewin. With therefrom a fine fish, hoisting it with expert hands, whilst Simon laughed and applauded. Every one, except the last one, he put back as carefully as he had taken it out. That was the way it was told to me, amid much laughter, for Glyn was a tipsy man, excellent in the water, but not very good at the gates and stiles and among the brambles. Never once did I see Glyn on that river myself, but then his was a very private calling.

In June, however, Dolgwili played host to a most delightful Englishman and his wife, a retired public schoolmaster or some such, most gentle in manner and informed on many subjects. They, and I, were much delayed one Monday morning by huge and continuous rain. It had fallen all through the latter half of Sunday, most of the night – we had been awakened by it – and by mid-morning our guests had decided that it looked like continuing all through their week of holiday and so they would dash down to their car and go off exploring bits of Cardigan Bay. Shortly after that, I too dashed down to get the Renault out, being in dire need of three-inch nails from the town. As I came from the farm on to the disused railway track, I saw a sort of man shape ahead of me. The shape was squat; it seemed to me he must be a cripple. As I came nearer I saw indeed that he was kneeling. It was still raining most heavily and, probably because of the original impression that he was crippled, I was all set to pass by rather than to stop and stare.

"Good morning," I said, en passant.

"Good morning," the wet man answered, apparently content for me to be merely en passant.

I stopped and stared. He had no hat; his face was white and shiny with rain. His jacket collar was turned up pathetically as though it might keep off some of that lashing rain, and his pale hand clutched at a rusty wire in the fence as if to steady him before he crumpled. The other hand still contrived a nominal grasp about a fishing rod and nearby rested his creel. He wore an ordinary, somewhat shabby suit, which was sodden, and in fact he gave the impression of a drowned man. He shook violently.

"Are you all right?" I asked inanely.

"No. I've broken my leg."

I began automatically to try to help, bending down to him whilst all the time normality splintered about me. Where did he live, where could I take him, I started to ask, and could he get to my car.

"I'll have to take you to hospital, man."

"I'll have to change my trousers first." he protested and we looked at each other like we were two lost men.

Sense returned. I took off my Barbour and spread it over his head and shoulders.

"Stay still now. I'm going back home to phone for an ambulance."

I ran all the way up the hill and flung into the kitchen.

"I've found an angler with a broken leg down on the track. He's wet through and frozen. Can you make him a hot drink whilst I phone?"

The rain had no mercy. We hurried back with a thermos full of hot Ovaltine and whisky or something and his hand shook so that he could not get the cup to his mouth. Phyl took the cup and fed him. Some ran out of the corner of his mouth. One wellington had come off; inside them he wore only small thin hosiery. He was still kneeling painfully on a small scree of railway-bed stones and we were still towelling him and feeding him hot drink when the first ambulance man came running up the track.

"Glyn, what are you doing here?" the young man demanded, almost immediately turning back to fetch, with his colleague, a stretcher.

In those few minutes when his nightmare was ending, as he drank and kept telling his thanks, he let fall a few of the bare bones of his story. He had gone up the river fishing yesterday afternoon and had turned for home when the rain turned heavy. He had been crossing the railway line almost two miles up from Dolgwili when he had slipped – nothing is quite as slippery as a wet railway sleeper – and broken his leg on the actual rail.

It is a lonely place up there. Simon used to say it was queer, too. One of his dogs, Urchin probably, had turned tail and run all the way home from there. The railway track and the river snake side by side through rocky gorges, bending for ever along the backsides of small farms, a place for occasional anglers, poachers, walkers and lovers. That Sunday, perhaps being Sunday and it being West Wales, came neither angler, poacher, walker or lover. Glyn began to drag himself along the railway back towards help.

They wrapped him up, gently laid him on the stretcher and carted him away. He would never forget the blessed woman, he was still averring as he went and when I followed him down to the white ambulance the driver turned to me and said, "He says you're to have the fish in his basket, and please would you look after his rod until he comes for it."

"Of course. You know him, do you?"

"Oh, yes. Everyone knows Glyn. He works at the hospital anyway."

The ambulance rushed away and I went back for the rod and basket. When I opened it, six quite small brown trout shone silver on a grass bier within. I examined the rod and tackle; by any standard it was cheap and elementary, and certainly not professional.

I thought about his night, and his journey. The railway track was the only level ground all the way, falling immediately from the grey stone bed down to the river to one side and climbing even more sharply up cliff and bank the other. Everywhere was lush with bramble, heather, grass or scrub undergrowth; everywhere was sodden. He would have had no choice but to drag himself from slippery sleeper to slippery sleeper and across the bruising trackbed stones, and all in belting, spiteful rain.

When that evening Simon came with his dogs, I related the morning's happenings and when he saw the rod and basket, he recognised his tipsy man immediately.

When that evening our English guests arrived home from Cardigan Bay, I related the morning's happenings and – "But we saw him," they exclaimed. "As we went out, just as we crossed the line, we looked up that way and we thought he was a workman doing something to the track . . . "

"But did he not call out to you? Not ask for help?"

"No, he didn't. I thought it seemed a bit strange to be working in all that rain all by himself."

A very private man. Out of curiosity I walked the dogs back up the track later and I discovered where he must have spent that wretched night. It had virtually no shelter. Even with me he seemed content to ask no help; perhaps he intended to drag himself over the bridge and across on to the road and up the road a couple of further miles to the village.

Later, when he had recovered, he sent Dolgwili – we were only Dolgwili to many, for our name was always a difficulty to them – two bottles of brandy with ever more thanks, and then, by Simon, a request to go and see him. He still was not properly mended. I sat and talked with him in his kitchen – he lived alone, of course – whilst his Jack Russell eyed my Englishness with some distrust and his master told me of some of his poaching exploits. Alas, my Englishness and his Welshness made poor fireside conversationalists; much of what he said I could not catch, so quiet did he seem.

Glyn was the archetypal Simon's man, and so was Dai.

"Down there," Dai Sheep instructed with an arthritic forefinger and pale eye as we sped towards the market.

As my trafficator blinked, I braked violently and the taxi that had been coming straight at me fast hooted angrily and swerved past my nose whilst Dai swayed and our poor sheep surged in close-packed unison in the back.

"Can't go down there, Dai. One-way street."

Oh Christ, he said in Welsh. "Down there then," he said in a minute. It was his town, not mine.

"No, that's one-way too."

Oh Christ, he said in Welsh; I think.

"I'll have to go round through the middle of the town now."

We arrived in the one-way system. People stared at our load. Dai kept wiping his nose on the back of his hand. He was old. I was aging fast. He was muttering to himself by now and getting thirsty, no doubt.

Inevitably we stumbled upon the market eventually, to join the queue of cars with trailers, tractors with boxes, Land-Rovers, vans, all loaded with sacrificial lambs. Somehow Dai Sheep wrenched the door open and fell rheumaticky out on to the summer concrete, scratching it noticeably with his indestructible bones before gathering himself to stand sniffing the air for the direction of ale. He had had enough of motorcars. He pointed vaguely to the lamb pens, or to the Drovers' Arms, or kingdom come and stumped away on his thumbstick.

Smallholders come in muddy wellingtons and stubbled chins. Farmers come in ties, nylon smocks and scrubbed wellingtons. Butchers come in good cloth and fine leather, with expectant smiles upon their well-shaved faces. The auctioneer discreetly comes in an amalgam of all. There is much pleasantry, many jokes in Welsh between him and the crowding buyers and sellers.

Our lambs weighed in at thirty-two and eventually sold for thirteen pounds twenty pence, helped thereto, although only marginally, by a single winked bid from somebody in the crowd whose face I suddenly recognise as Dai Sheep's.

"How are you getting home, Dai?" I searched him out when the tide of men had flowed on along the selling pens, and asked. "Because I'm going home now."

"No, no, no, no, no, you can't go home. You have to be here when they sell your two old ewes, see?"

"Why?" It was the last thing I wanted. Scratcher had lost her lambs two years in succession and Disappointment never had been in lamb, so they had to go, but I was cowardly enough not to want to see their going.

"Why, to see them sold. To be there."

I hesitated. Was it merely that he himself did not want to go yet, wanted a lift home later? Was it protocol, was it really necessary? Well, he was old, and he was Simon's friend; I sighed, and stayed.

"Come on, Dai, I'll buy you a drink."

The pub was cheerless, and so was the ale. Dai Sheep and I did a bit more of our cross-talk act and drifted back to the lamb sale where Dai fell among cronies. The butchers laughed a lot, the auctioneer pointed and beat with his stick, talking fast, the paintsticks daubed yet more coloured hieroglyphs across the fleeces of the doomed lambs, a man punched holes in their ears, producing smears of blood and the crowd milled and pushed ceaselessly. The whole business made me dispirited.

For two hours I danced irksome attendance on Dai Sheep, waiting for them to get round to selling those few forsaken old ewes. I wandered through the neglected corrugated-tin buildings, read old advertisements, wasted time inspecting wares in agricultural shops, viewing the passing show with distaste, avoiding the area where waited Scratcher and Disappointment until the last minute. Elsewhere huge transporters waited with great maws open to receive the running lambs now being chivvied and loaded by shouting men.

Scratcher produced a single bid of four pounds and when the auctioneer's eye queried me, I nodded glumly and let her go out of my life with only passing regret. Disappointment was a much larger sheep and as soon as she was offered a wizened old man urgently demanded of me what was wrong with her and at the same time as I was repeating – for no Welshman ever understood me the first time – that we were never able to get her into lamb, the auctioneer's bidding involved some sort of private pantomime between two burly young butchers and old Dai Sheep during which the bidding mounted a bit and suddenly the auctioneer's eye was on me once more gently asking was ten pounds all right.

"Okay," I said, no less glumly, and turned away ready to get hold of Dai and back to Dolgwili. I had just been converted to vegetarianism.

"Are you ready now then, Dai?"

"Er . . . Where's your motor?"

I pointed. "On the carpark."

"Well, er . . . " He looked about him exasperatingly. "I . . . Yes. But I've got a bit o' business to attend to first, your worship."

He examined a screwed-up piece of official paper, moving it close to those pale eyes. He is one hundred and seventeen years old; he can still see to read without glasses.

"What business?" I demanded shortly.

Dai points. "Sheep," he bellows impatiently. Is there any other sort of business then?

"What sheep, for goodness sake?"

Just as impatiently he points and shakes his thumbstick towards Scratcher and Disappointment. What does he mean?

"*You* bought them?"

He nods, expressionless, and looks about him. Perhaps he is wondering which of all these mighty cattle-trucks he should choose to convey his sheep.

"Get your motor over from the carpark, see," he turns back and instructs me. "Put them back in that shelter of yours and I'll get Simon or my son to come and fetch them tomorrow, see."

I observed his toothless mouth, his thin grimacing face and his famous hat which he raises frequently for scratching purposes, and I said, "You damned old man."

I thought, as I drove the Renault too briskly from the carpark over to the iron pens where the old ewes had spent as miserable a day as I myself had, of all the splendid things I could have been doing among my roses and shrubs all these market hours, of how much better it was to be alone among flowers than to be in the company of men. I was, in fact, seething.

Dai Sheep beckoned to one of his minions who hoisted Scratcher and Disappointment back into the tumbrel, and then, wordlessly, I drove Dai back to the village and the two old ewes back to Dolgwili. Lastly I made a mental note that I must warn Simon immediately – and indeed the rest of the innocent universe – not to be buying any breeding ewes from Dai Sheep at about, say, seventeen pounds, for the next few days.

18 Saluki

Heat occupied the land. The grass scorched; ponies, donkey and sheep stood sulking in shade all day, worried remorselessly by flies, tails flicking, eyes closed in patient misery, waiting for late evening coolness before emerging to wander listlessly up the hill at last to graze. Roses and raspberries grew thinly, beetroot leaves collapsed, young cabbages grew skeletonised by hordes of munching caterpillars. The earth shrivelled, the thin river starved almost to death and displayed its grey-white rocks like bleached bones. The glare of the sun pinned me mercilessly and filled my head with floating pain, so that the only work done daily was the endless carting of water to the panting sheep. The conservatory, clouded in whitewash and shrouded in shades, with its door and all six windows open, stifled under one hundred and ten degrees by day and eighty by night. The tomatoes sucked water continuously to stay alive and only the epiphyllums sang in glory to the sun. Out on the pastures, the thistles seeded and the bracken grew even taller, whilst us two sheep-farmers dressed ourselves as Arabs and sat helplessly clutching glasses of blackcurrant syrup and dreamed of rain.

Perhaps, that one day there at Dolgwili, the glaring sun achieved an extra penetrating power sufficient to bamboozle first the cactus and then the saluki into thinking themselves back respectively in the arid heat of Arizona and of Persia.

At our dinner table in the cool little room at the end of the cottage, which the evening shade of the four great conifers rendered cellar-like in temperature, our guests for the week sat happily exchanging reports on their different days at the coast or up in the hills. As a sort of appetiser, I first presented to them a cactus from the conservatory, just starting to open. The spiky green thing, lumpen so long, had suddenly transformed itself with a strange long bud which had stretched upwards rapidly to a point six inches tall.

"There we are. You can sit and watch it open as you eat. It is quite an extraordinary flower. Quite fantastic."

The lovely lady looked at me out of a glitter of ear-rings and careful grooming, seducing me into adding, "And as the petals open," my cupped hands formed a fine corolla and my fingers slowly opened as petals, "the flower utters a sound like brass . . . baigh, paigh, paigh, paigh . . . like that, a sort of fanfare."

Perhaps that great sun had reached in, too, to some long-buried ancestral memory within the lovely lady's head, for she blossomed herself into a sweet smile of pleasure as she granted me her total belief in the possibility.

"Will it really?" she exclaimed, clapping her bejewelled hands together in delight. "Ough, how lovely."

And although her companions greeted my light-hearted promise of natural psychedelia with laughing disbelief, I thought I noticed that she herself wore a slightly expectant look as I waited on their table. At a rate just tantalisingly slower than visible movement, the gorgeous flower pushed its pale petals open from the green sheath of bud to become, within no more than a score of minutes, a perfect and utterly lovely bloom, the palest of lemons touched with hints of green. To gaze down into its throat, as we all did, was to gaze at indescribable and unsullied beauty. Amazement abounded.

Just once my smiling eye latched on hers. Her head might have been slightly to one side, her tongue was tantalising her teeth in a delicate smile. We exchanged tiny nods, for she and I had privately heard across the black silences of our mind that brassy fanfare out of the long cornet of the perfect bloom, a salute to utter beauty. I put their black coffees before them and returned to the kitchen.

"Thank goodness that's over," my sweltering cook-shepherdess muttered as she wandered away from her enslaving Aga to seek the slightly cooler air outside. Sam-Boxer immediately lugged himself in, despite the summer heat, to replace her at the stove,

"Where's the other one then?" I asked, looking about me middle-agedly.

146

Phyl shrugged. "Caramelle? She *was* here . . . In the conservatory."

At the same second a frantic Trudi came running past, crying, "Caramelle is chasing the sheep."

I dashed out swearing. As we ran along the terrace yelling, I could see traces of the flock galloping in panic up the slope but I noticed that a single lamb was tearing off down the slope towards the railway fence. Caramelle was bounding after it, only yards behind.

"Caramelle," I roared threateningly from the bottom of my throat. "Caramelle, come here, Caramelle."

As Trudi and I went through the gate into Sheepdip, the lamb dived through the fence and disappeared down the bank into the trees and scrub of the railway embankment. The saluki bounded about excitedly by the fence, temporarily frustrated.

"Caramelle, come here, blast you."

Caramelle slipped through after the lamb. I did my gentle gallop towards the fence, calling desperately. Within some calm region of my head, even as I galloped, I knew that I might be approaching the death of the saluki. Beyond the fence the bank was hazardously steep. Gasping at breath, I slithered down from ash to sycamore, searching among the thick weeds and bushes for my damned dog, whilst somewhere above me Trudi screamed angrily, "Caramelle . . . Caramelle . . . Caramelle!"

"Can you see her? You stay up there in case they break back. Can you see them?"

"There they are."

I followed her pointing hand. I could just see the saluki and she seemed to be standing over the lamb. The lamb did not seem to be moving. Both were half-hidden by undergrowth.

"Caramelle, leave it," I shouted with my angriest voice. "Caramelle, get away."

As I slid and fell down towards them I could see no blood on the jaws of the dog but her eyes were wild. I came to them shouting and waving my arms alarmingly and the dog ran off into the trees. The lamb did not move from its slumped position on the grass and weeds.

"Watch her," I yelled to Trudi. "She's going towards the bridge."

I knelt over the lamb. There was no blood. It was alive, but totally immobile, paralysed by fear. I picked up its limp body and examined it, but it seemed quite uninjured; it simply could not stand. Carrying it, I began to climb back upwards, with neither breath nor handhold.

When eventually I regained the meadow level, I had nothing left except the lamb in my arms. I sank to the ground whilst Trudi came towards me, the saluki on a handkerchief leash, her tongue lolling and her eyes still sparkling from the excitement of the chase.

"Is it all right?"

"I think so," I gasped and sat quite still on the grass, as still as the lamb in my arms. "Take Caramelle home, will you, please."

I sat and watched my daughter and my dog go. I sat there a long time in the shade beneath the sycamores whilst my breath and strength returned. I was thinking of Caramelle.

She was my dog. I had chosen her chocolate beauty when she was but twelve weeks old and that day she had been carsick on my lap all the way home, bonding us to each other. Almost from the start she had been unsuitable, but I had made my choice and would keep to it. Independent, fragile, nervous, aloof, aristocratic, jealous, difficult, but so beautiful . . .

She had been condemned to die before this. An accident-prone puppyhood ended with her suffering demodectic mange. Her thin, bald body became ravaged by suppurating sores, and every ten days for many weeks we had carried her strugglingly, leggily upstairs to shampoo her with medicaments in the bath, where she stood miserably shaking, slipping and crying whilst we tried to soothe her and comfort her for the ten minutes necessary for the treatment. The loathsome sores took no notice of the treatment; the brown hair continued to fall. Other treatments, other failures, and all the time the young saluki had to be isolated from the boxers and the terriers and, to some extent, the home. For ten months we struggled. The period coincided also with financial struggle and finally with physical struggle as illness laid me low enough to be carted off to hospital.

"You cannot possibly cope with Caramelle as well as all the chickens, goats, ponies and dogs by yourself. You must have her put down," I feebly commanded from the stretcher, as it were; the skids were under the lot of us then.

"We'll see."

She had found a quack remedy, one advertised for such a diversity of canine complaints and for so little cost that it could not possibly be taken seriously. Yet, from that day when I entered hospital, when the old-fashioned ointment was rubbed into that appalling body, Caramelle began to improve just as miraculously as one always must hope a quack and cheap medicine will. The sores began to dry and scale off and soon the first hair to grow. What a joy that was, and how thereafter could we not all mean much more to each other. All that was years ago. Now, as I sat still nursing the lamb, much longer than I need, the dog's fate was in the balance once again.

Times out of number I had worried that this would happen, not so much to our own sheep, but to neighbours' sheep. Before our own flock was ever dreamed of, Caramelle and Sam-Boxer used to chase free as the wind morning and evening all across the Dolgwili slopes, whilst I used to watch them keen-eyed with anxiety lest they see a bolting rabbit or hare which they might chase clean through our thin hedges on to Thomas' land. The worst nightmare there was, then, was to imagine saluki and boxer gone off on a sheep-worrying spree all through West Wales.

Well, they never had, till now. But I had always known that the saluki was a mistake. I chose the breed, as I chose the dog, for beauty alone; it never had been, within the family, a comfortable choice. And the dog, to a quite ridiculous degree, was so much like the man, especially in these latter days of so much self-knowledge. Both of us were temperamentally alike – anxious, highly-strung perhaps, suspicious of strangers, always slightly edgy. She was a loner; so was I. She yelped at the slightest hurt or even threat of hurt, and once, in an astonishingly revealing moment late in married life, I had quite incidentally heard my wife confide to my daughter, "Yes, he was a beautiful mover."

She had never told me. In all those years. But if I did not know that I had been a beautiful mover, I had always known that I was a joyous mover. And so was Caramelle. Both of us gloried in our bounding stride across green turf, and now that age had shortened my stride and sheep had limited the saluki's, we had come to enjoy each other's company more soberly with

149

walks along the river or the sands of the estuary or anywhere free of the temptation of fleeing sheep. Evidently she had been in the conservatory when she had seen a flight of sheep streaming from our garden paddock through the gateway into Sheepdip, and although there were theoretical defences to keep the saluki both from leaving the conservatory and from the garden paddock, this evening they were obviously not enough. The sheep had suddenly become the gazelle she had been bred to hunt across the deserts of the Middle East where once Arabs valued saluki above women and honoured them not as a dog at all but as an animal prized above all else. It was not Caramelle's fault that she saw the sheep as gazelle and gave chase; it was mine own.

"Are you all right, Dad?"

I opened my eyes and sat up. Wife and daughter waited across at the gate, and behind them hovered our guests taking their evening stroll.

"Yes, thanks," I answered tiredly, placing the lamb down to see if it now could stand. It stood. It baa-ed. Its mother answered from among the now placidly grazing flock up on high ground and in a moment the unscathed lamb began to walk and then to run off towards the sound.

I stood up just as well as the lamb had done. Phyl looked me in the eye significantly as I came near her; there was, I felt sure, something challenging in her look. She had never quite been able to forgive the saluki for not being one of her beloved boxers.

I smiled cagily back and I said, "We'll see." Yet even then, in that moment, I knew that if I had not now decided that Caramelle must be put down for sheep-chasing, then by tomorrow such a decision would be even more remote. After all, she had not actually killed or even harmed the lamb. After all, it was not her fault. And she was so active, so lovely. And she was my dog.

19 Don't look at me like that

We arrived at the demonstration ten minutes early, and thus defined ourselves as rookies. We were rarin' for sheep. One of half-a-dozen official young men emblazoned with badges and burdened by leaflets came at us full of smiling and soft-skinned charm as though after an order.

"You are sheep-farmers?" he asked, handing us leaflets on scab and on sheep-dipping.

"Well," we temporised, for the description flattered us over-much. "More or less. In a small way."

"Have you got your own dip?" he teased, casting a salesman-arm vaguely about him. About fifty all-of-a-kind, all-of-a-size Suffolks stood immaculately penned adjacent to a newly-built dip, a glorious thing to our deprived eyes, of shining tubular steel and concrete, square and impregnable, without an old piece of string or rusty wire in sight. Every gate was correctly hung, every pen thoughtfully placed, every draining slope utterly right. We drooled.

"No. Well, that is to say, we *do* have an old collier's cart, rusty and semi-derelict, abandoned from the old days when they used to dip, standing in one corner of a field. We did think of persuading a neighbour who has five smelly sons to come over and sink it in the ground so we could both use it for dipping."

"It might do. How many sheep have you got altogether?"

Strict truth always deserts me at times like this. I added about twelve per cent for aggrandisement and remembered not to mention Sam-Ram or our neighbour's scruffy eleven which had not been shorn this summer.

"Well," said the young adviser, kindly enough, "if I were you

151

I would not bother. It's hardly worth installing a dip for so few sheep." It was the story of our sheep-farming life. "I feel sure you can find a neighbour who will let you use his."

"If I had a neighbour with a set-up like this one, I would not dream of considering anything else. But I haven't."

Actually we were so sheep-crazy now we wanted to do everything ourselves, even dipping, and we wandered about the farm where the demonstration soon was to begin more covetously than was right. The farmer was his own shepherd, looked to me as if he knew every last little thing about sheep, which all looked superb, and was obviously a most happy man. Admittedly his farmhouse looked uninterestingly drab compared to the whitewashed stone and tarred roof of his farm-buildings, but this was standard – just as standard as the lateness of all the van-driving, Land-Rover-driving farmers who now suddenly arrived and swarmed round for the demonstration.

Scab, that loathsome thing, had reared its ugly head once more and so dipping was compulsory again, the speeches reminded us, but dipping was a dangerous business and needed to be done correctly. As follows:

The poor farmer-shepherd started togging himself up like an astronaut – black protective clothing, gloves, face-visor. He opened the chemical dip as if it were high explosive. Clean running water must always be on hand, no dip must be allowed near the eyes or the mouth – of the dipper, that is, not the dipped. A young helper yanked a lovely ewe from the waiting-pen, kneed her directly down into the fiercesome bath where the shepherd pushed her head under briefly with his special crook and kept her swimming around in the creosotic brew for a measured minute, which seemed like half-an-hour. As soon as the dipped ewe found the exit ramp and climbed up to stand dripping, another ewe was dumped in to take its place. It was all so efficient, so easy, so enviable, that on the way home we resolved again to win the pools.

We did have until mid-November to decide how we would get our flock dipped, but we were fairly certain no neighbour of ours had the Rolls-Royce type dip. Mr. Thomas almost certainly would dip his own, and perhaps if Phyl could engineer a meeting with him, smile appealingly and talk very slowly and distinctly, then she might get him to dip our sheep with his – if, that is, we could think of a practical way of getting them up the track and across his farm. Since he had already done our shearing, however, it was unthinkable that he could also undertake our dipping. No; we would have to dig a big hole and make our own dip.

The days remained hot, the ground was very hard. The hole never got into the ground. Very soon Simon began to talk vaguely of a new dip constructed on a farm near the village. Perhaps, he began to say vaguely, ours could go there. Well, perhaps. But the accumulation of a kind of delayed homesickness for real English villages, real English gardens and stately homes and what we thought of as 'the English quality' pushed makeshift plans for sheep-dipping right out of our mind. Yes, all right, we would see to it, and yes, all right, Dolgwili's acres were splendid, and Wales had been kind and generous to us, but for now we wanted England. The Welsh towns and villages seemed poor and ugly; everything seemed rough here. Ah, to be in England . . .

For days we wandered the Cotswolds almost to our hearts' content, finding places that were the England of the calendar photographs, town squares created for film-makers, age-old villages with names created by poets, and houses out of *Country Life* advertisements. Yet very soon Dolgwili was calling us back, for Chipping Campden and Stanway, lovely though they were, had no personal meaning; we were as keen to get back as we had been keen to leave.

We had scarcely greeted the grandchildren and taken a peep at the flock when the kitchen door flung open and Simon entered with, "I'm helping old Dai and Mulligan dip their sheep up at Dan-yr-allt. Do you want your flock done?"

"When?" I demanded sharply.

"Now."

I must have frowned. "But we haven't informed the office. You have to fill forms in and give three clear days' notice so they can send an inspector out."

We glared at each other. He grinned. He was getting very Welsh. It was a Sunday; what inspector would come out on a Sunday anyway?

"What about Dai and Mulligan, have they filled their forms in?"

Simon shrugged. It was of no importance.

"But how will we get the sheep up there?"

"Drive them up the road. There's not much traffic. They can all go together, all three lots."

He watched me hesitating, saw my disapproval of all this spur-of-the-moment stuff. "The boys are waiting," he said.

"All right," I agreed with no enthusiasm and having no idea how else I could get them dipped. "But listen. I want them dipped properly. Make sure each one gets a full minute or else it's time wasted and make sure each one is ducked. You do know what you're taking on, don't you? You know it's danger-ous stuff, this dip?"

I nattered at him a bit, over-fatherly no doubt; after all the years of rearing him, I would rather he did not come to a sticky end in a sheepdip.

Going, he suddenly said, a little peevedly, "It *was* all ar-ranged, wasn't it?"

This was the way it was always, in all things, indefinite, foggy. Once I used to think that fog was something which enveloped only us and our Dolgwili sheep, a fog of ignorance and inexperience unknown to other folk. Gradually, however, I had come to suspect that the fog was general. Nobody knew much about anything. Even those admirable characters around the dipping demonstration, farmers who in appearance seemed knowledgeable and confident, had shown themselves by their questions to be groping, just as we were, in fog. It shrouded all the answers, too, and all the arrangements. Nobody ever quite knew for certain anymore. Detailed programmes, keenly-thought-out plans, fine ideas, excellent organisation, all the splendid parade of man's ingenuity and care simply got bogged down in the fog. Nobody quite knew where they were going, nor why. The map which had been so abundantly clear at the start of any journey could no longer be trusted. You started out with a song in your heart and fully confident, but soon the signposts had been twisted round, unmarked side-roads appeared and first turnings on the right always ended up in somebody's farmyard.

"No, no, the other way, see? Back to the road, turn left and go straight over the square and past the post office and you'll see a pub and there we are, see. You cannot miss it."

It took us two years to discover that 'the square' means crossroads, that almost every other village is called Velindre, that north shifts quite often, that right means left.

One guest family that summer had sought, at my suggestion, Llyn Brianne, a great new reservoir set in beauty up in the hills. The poor chap drove two hundred miles that day, following maps, asking directions, coming to temporary signs which said 'No Through Road', trying this way, trying that way; he never did find it. A great reservoir seven miles long, gone, lost in fog. Guidebooks beguile you to seek waterfalls that are unfindable. Great adverts announce Sale, Sale, Sale, but do not reveal where. Even gardens open to the public coyly do not name themselves. At the agricultural shows huge marquees stand grey and signless, glorying in anonymity. Dolgwili had three addresses that we knew of. Uncle Dai had delivered mail to us from one village post office, we lived closer to another village post office, and officially, it now seems, our address is that of a third village, although we get mail delivered it is advisable not to have any of those villages mentioned on the envelope but to have the town mentioned instead. Fog, fog, it is fog all the way. You never know when the builders, the shearers, the dippers or anybody else is coming, nothing is as it should be, nobody knows where they are at.

Or somebody looking just like your own son looms up out of the fog at your door, on the wrong day, at the wrong time to announce that everything is waiting and arranged, and whilst he takes your sheep and drives them up the road, you stand there wondering at your own senility, that so vital an event as the dipping of your own sheep could arrive without you knowing a thing about it.

I did know where Dan-yr-allt was. I had met the English farmer. Having allowed Simon and the flock sufficient time, I jumped in the Renault and drove up to watch proceedings. I mean, I knew they were there, somewhere. It was hot, it was hilly. I trudged about the fields, the tracks, the paths looking for a flock of sheep, in the charge of about eight men, and a sheepdip. Of all, there was no sign. All, like Llyn Brianne, had disappeared, in the foggy, foggy dew of my mind. Where in heavens' name could that lot be? I trudged on and never a track-sign did I discover, never a 'baa' did I hear, until, lost, sheepishly, I went and knocked at the farmhouse door to ask, "Please, where is the sheepdip."

No answer came. I trudged around some more, contemplated going home, beaten, contemplated my own imbecility, my own mental and physical exhaustion. A different Dan-yr-allt perhaps? Farm names repeat themselves as often as village names. Dan-yr-allt, Velindre, perhaps? Perhaps if I returned to Dolgwili I would find the sheep all back there. I was just driving away in a mood of baffled anger when a Land-Rover met me at the entrance; it was the return of the English farmer and his wife from church. Perhaps I had failed to erase all my baffled anger from my query, or perhaps they noticed my near exhaustion, but they took me into the farm kitchen, sat me down and gently plied me with reviving sherry and coffee and unhurriedly assured me that all would be well and that we would all drive up there in a minute or five and see.

Soon we began to ascend a steep track in a great grey tank of a Land-Rover, whose power taunted me into dispirited comparison with our puny little car and all the rest of the puny little aids we had at Dolgwili. I felt vastly inferior, vastly grateful and vastly relieved. Down the land towards us now came my ragamuffin crew of sheepdippers, laughing and shouting as they convoyed our motley flock of dripping sheep away from the dip and down towards the road back to home. They assured me most earnestly that a thoroughly good dipping had been accomplished, however officially unwitnessed, and as I observed their rough old clothes and boots and sticks, I remembered all the meticulous dressing-up and care displayed at the ministry demonstration. Ah Wales, I thought, sighing. I thanked the kindly farmer, gratefully accepted his suggested terms of a couple of quid for church funds and then slunk home in the Renault to await the return of the flock.

Sam-Ram came home a poor last, individually escorted, and quite done in. He walked slowly and I took him over from Mulligan and accompanied him up to the stable. We had, I realised, got it all wrong. The long walk up the road on a warm day, the stress of the dipping, and then the long walk home on those poor old feet had all been too much. We cosseted him once again in the stable with new bedding, clean water and food and I sat and explained how much better things would be if a son would only leave a father to do it all his way. He seemed to understand, but wherever I looked question marks hung. Having nursed Sam back from the valley of the shadow in winter, it seemed palpably absurd to hit him over the head in late summer, although there could now be no escaping the fact that his days were numbered. His heir, Samson, had already been sold to Mulligan, for we had decided instead to keep our ewe lambs as flock replacements, and they, of course, were Samson's sisters. We were well on into autumn and Sam had already tupped the first ewes, but after the dipping he was markedly unreliable; how could we possibly leave the whole future of the flock to this poor old clapped-out gentleman?

The suggestion that we go in for pedigree breeding had been mulling around in our collective mind for weeks, lost in that indecision which seemed to be the hallmark of our husband-and-wife system of management, or non-management. Now however the suggestion – that we go in properly for rearing pedigree lambs, preferably Suffolks – came stampeding out into the open. We rushed about to the ram sales, we talked and sought advice, we did our sums and finally an advert in the local paper sent us off into Cardiganshire bubbling over with enthusiasm to buy Suffolks.

The young man wore a lean and hungry look, as well he should, for he filled his long day by milking eighty cows, shepherding his father's flock of Speckleface, breeding and showing ponies, raising his own young family and, for a hobby, running a small flock of pedigree Suffolks. When he confessed, or boasted, that he had just been up to Scotland, for my dear old Suffolk apparently has no sheep now, to buy a pedigree ram for four hundred pounds, I knew we had moved up in the sheep-world. Blandly he assured us that Suffolks were much easier to handle than Speckleface, that they were docile and easily managed, that they lambed much earlier and had to be 'done' well but that for ever there would be a very firm demand for Suffolk rams, and indeed so many other blandishments that it was like taking bellows to the fire that burnt within us already. We gladly paid him seventy-five pounds each for three huge ewes and forty-five pounds for a ram lamb to take over from Sam.

This he did with immediate relish, and although by the time Golden Boy had arrived, Sam-Ram had struggled to his feet and shewn willing to return to duty, when he saw Golden Boy's energy of attack, then understandably he retired back to the cosiness of the shelter. Very soon he was back to where he had been last winter, in the valley of the shadow of death. This time the worm-remedy had no effect.

I fed him regularly, visited him often, talked of the weather and was there anything he fancied . . . grapes, perhaps? What do you do? I pondered for a fortnight over him until I could see no alternative. I phoned a man who advertised in the local paper for 'fallen beasts', and when he sounded friendly and understanding, I brought the little Renault up to near the stable, lifted Sam-Ram on to straw in the back and drove him down the road to the knacker's yard.

He looked a much-scrubbed man. He had emerged from a collection of tin buildings with concrete floors, all of which too seemed much scrubbed. I noticed that among the buildings and in grassy places he had stood a collection of rather classic-looking white marble busts, many of them on columns, as if to add class to the place.

He apologised, saying, "I really must go indoors and get my dinner. It's out on the plate and I've been working flat out since six this morning and I'm quite famished. Just put him over there. Let him graze in the orchard and I'll come out soon and give him a pill."

The man went indoors. I opened the back of the Renault. Sam was very weak and I lifted his nearly weightless body down without a struggle. He had begun to scour badly again. He followed me over and began to nibble at the nice grass without enthusiasm. He looked up at me, for nuts perhaps.

"'Bye, Sam," I said. "And thank you."

I jumped into the car and drove off without a backward glance.

Well?

Don't *look* at me like that . . .

20 I dreamt of haystacks

"I don't think they're a good idea," I reflected as we leaned on the gate together, gazing at sheep. It had stopped raining for half-an-hour. It was the year's nadir, mid-January, a slow time of despondency. Nothing was a good idea now, except hibernation.

"Paddocks?" she queried, in some surprise.

"Suffolks," I corrected, with undertones of rude crosstalk.

"Why ever not?"

"Well, look at them. Damned great hungry things."

Phyl pouted. We had only just fed them with what we thought of as generous helpings of expensive concentrates, yet they glared and blared at us still.

"Why can't they clear off up into the wood with the Speckles and stuff themselves with nourishing ivy and blackberry leaves and bark. Every time they see me they look straight at me and start shouting about their blasted young kicking in the womb, always demanding more and more and always the hard stuff. They won't hunt and they won't look at hay and – "

"Well," she defended them reasonably enough. "They're close to lambing."

The truth was, they worried me. They really were such huge things. Damned great mothers filled with damned great lambs, I reckoned, and so with damned great appetites. I was still haunted by what young Davies-Lean-and-Hungry had said when he had delivered the Suffolks to us.

"Oh yes," he had thrown out casually, "you've got to do them well, of course."

It was an obvious enough thing to say, but it caused me to mooch about the farm examining what remained of the grass, to sit for hours with textbooks – all of which seemed either to have been written in the 1930s or for vets and agricultural statisticians – and to wake at night wondering how we would get by. How could we do them well? We had no grass, no rape, no turnips, no cabbage, hardly any hay, and anyway this pedigree lot turned up their great black noses at everything except nuts. To go into the flock now with a bag of nuts was almost to take your life in your hands, for the Suffolks, much less shy and much more bulky, came straight for you, brushing wetly in front of you to halt you, and then pushing their heads into you or beginning to climb up you with booted hooves as big, strong and foot-rot-free as though they wore army boots. Golden Boy, shy at first, soon began to butt us playfully at about mid-thigh, and all four Suffolks, even at dusk when I traipsed around to check and shut-up, were still loitering and calling out at me for grub when all the Speckles were decently tucked up on the highest ground or in the wood.

"How much would that electric sheep-netting cost?" she started again.

"Too much," I said automatically and too brusquely. Well, it had started to rain again, and the ponies had come chasing down the slope, excited at the possibility of more carrots and crusts, and we had hardly any hay and it would be months yet before grass grew. The boss chestnut pony, Fleur, was in foal again probably, and she demanded preferential treatment. She bullied me as much as she bullied the other three ponies and the donkey; I always felt mean at times like this when I could offer her no more.

"We would have enough money for electric paddocks if we sold the ponies," Phyl said, starting a campaign as we went indoors. With good cause she had become apprehensive this winter of their wheeling bodies and tossing heads as they charged towards the mere glimpse of food. Nobody yet had actually been killed, but it tended to grow more like Waterloo daily as winter-hunger took hold. This, however, was the first mention I had heard of selling the ponies to buy paddocks.

Paddocks, I knew, would give her no peace now. Lately the sheer audacity of us, of all people, breeding pedigree Suffolk sheep had rather intoxicated her, and just as half a lifetime ago when we had stood watching the capers of Dai Sheep and his grand-daughter, she had lately been at my shoulder again whispering as seductively as any pipe-tobacco woman:

"Paddocks . . . " or,

"Pole barns . . . " or,

"Fertilisers . . . " or half a dozen other improvements or additions or fantasies.

The truth was that one thing invariably led to half-a-dozen others, and before you knew where you were, you were on a treadmill. Suffolks, so different a fold of sheep from our old Specklefaces, were leading her light-headedly towards a treadmill of improvements which only a year ago we would have considered areas of mystery inhabitable only by those we thought of as 'proper farmers'. Some of the fog was dispersing; hand-in-hand, we were pressing forward with a little more confidence. Small searchlights – books, radio and television programmes, the farming press – pointed our interest. One television searchlight had shown us a proper farmer folding sheep across turnips simply by carrying under his arm a curtain of plastic netting attached to its own lightweight fencing stakes, each of which he simply pushed into the ground. Within minutes a neat and sheep-proof electric fence fifty yards long stood before our very eyes. Could we not do that? We still had our old fencer unit from the goat days, didn't we? Wouldn't the garden be a lot safer using a fence like that? The mind began to boggle as it considered the changes such a fence would bring about. We could at last subdivide Dolgwili. Poor hedges would no longer matter, the unthinkable expense of a ring fence could remain unthinkable. Paddocks, it would be paddocks all the way now.

What on earth had got into the woman? How could the home-loving, fastidious housewife and mother, the wearer of lovely clothes, the flower-arranger and marvellous woman I had nurtured all these years suddenly now conceive all these agricultural ambitions proper only to farmers who employed shepherds, owned Land-Rovers and had hundreds of flat culti-vated acres? Paddocks, indeed!

Still, you could see, couldn't you, in the mind's eye all the advantages. These big, soft Suffolks and their preciously early lambs. You could just fence off their very own little paddock on the lawn. Move them around every few days. You could fence off the ewes with twins from those with singles. You could wean the lambs that way. Come summer, if ever it did, you could paddock a bit off for silage. Cut the May grass with the good old ride-on mower and make it into silage in bags. Have the trailer behind. By God, eh? Control. Control and management at last. Better stocking rates, better everything.

I was on the treadmill of ideas.

Flexinet paddocks, soil-testing and fertiliser advice from the ministry, liming, chemical spraying of the persistent patches of thistle across Lower Rough, and the nettle all along the margins . . . We could spray the bracken, too. We'd castrate all the male lambs – other than the precious Suffolks, of course – and we would ear-tag to identify progeny. We would inoculate, and we would put in our own dip and, by heavens, we would go crazy in the glorious outburst of sheep-farming razmataz. All this because we had bought three pure Suffolks and heard about electrified netting. We did not even know as yet if we could manage Suffolks.

Last autumn, on reconnaissance at a local ram sale, I had with suitable temerity approached a man who had been pointed out to me as being one of the best local breeders of Suffolk rams. He was just leading the best ram lamb that was on offer that day back from the sale-ring to a pen, prior, I assumed, to taking it back home, for he had just refused an offer of sixty-something pounds for it.

So as he led it on its halter across the cement I fell in beside him and asked, "How much do you think he would weigh when he was full-grown?"

"I've no idea," the farmer answered, pleasantly enough and indeed smiling and looking me in the eye to indicate, I supposed, that really he did not lift his rams that often.

"Well," I said, all ready to explain that my sheep-farming career so far had largely been taken up with lifting a ram who suffered from continuous foot-rot. I desisted, however, for it would have been an affront merely to suggest that such a magnificent animal as this one could ever succumb to such an unpleasant condition, and in any case by now a small ring of admiring farmers had gathered round the young ram. He was table-sized and table-shaped. His wide back had been trimmed, making it almost impossible not to touch. Hands and fingers gently reached out and felt him, as though in a spontaneous act of grace and blessing. He was all that poor old Sam-Ram should have been and indeed once may have been in that time before he came to Dolgwili. This one's fleece bloomed with well-being and care; his great black head and legs were thick and active. I found him almost irresistible and but that my brain was so doltishly obsessed with the dreadful weight of the fine creature I would have offered him seventy pounds then and there. After our homage, the breeder led him away, leaving me to estimate that the young ram could not possibly weigh much less than two hundredweight – and this was still a lamb. I could not see that he could possibly weigh in at less than three hundredweight in a year's time. How on earth would I ever manage a three-hundredweight ram? I'd never throw him; he would throw me.

Our own version of that splendid animal, Golden Boy, had neither the presence of the market ram nor even of our old Sam-Ram. He was as yet nowhere near two hundredweight, I felt sure, although the occasion had not yet arrived when I must lift or push or wrestle him. He was inclined to sulk and to be somewhat shifty-eyed, and the line of his back was a little too humpty to be handsome, but then he was, after all, a spur-of-the-moment and somewhat makeshift buy. Although the whole ram-replacing episode had taught both of us much, even to the extent of making us believe at last that we knew a good animal when we saw one, it was hardly conceivable to us that such top-class rams as the seventy-pounder pedigree one were ever destined to come to Dolgwili – not at least for the time being. No; we had graduated from Sam-Ram to Golden Boy, that was a satisfactory progress. He took over from Sam without a hiccup, donned the raddle-harness and mated his ewes eagerly, and daily as we fed and inspected the flock we noted with great relief that he never went out in all that mud without wearing his goloshes. After Sam's disastrous footwear, Golden Boy always looked immaculately shod.

It was just as well, for all the winter rains still had to be endured. It was the slow thin time of leaden days; the earth was always sodden. Mere showery days were good days, even when the showers were spiteful. It was the day-after-day, day-after-day rains that were so unendurable, that made hibernation the only attractive policy. Ditches filled and overflowed, the shingle of our pathways washed ever downwards towards the swollen mud-coloured river along ever deepening ruts. Water-bombs formed a yard below the surface, new springs started, everything ran with water. The pounding hooves of ponies and sheep poached the dead pastures in a search for the very grass which they themselves destroyed. Across the Steep, dark brown smears of dead bracken were beaten ever lower into the earth as the rains poured across the hillsides. Only in the black wood was there respite for the animals. On the stony thin soils of the hillside, ash and birch died young as rains and rivulets rinsed away the earth from their claw-hold and sent them slithering down towards the ultimately all-devouring river. It was a time of little pleasure for man or beast. Yet, even then, plants grew. The slow underturning of time and earth had known no halt, and nor had growth. Wherever you found a moment to peer, hazel and blackcurrant, clematis and burgeoning bulb, all things wore small green badges proclaiming that the age-old machinery of the seasons had not rusted completely but was still just ticking over, biding its time, ready for the long haul up into spring.

No month is as slow as February; its fewer days are the longest. They simply will not move. The shortest day has gone, the snowdrops are showing their whiteness and spring and the new grass simply cannot be that far off. Yet the long old days lug and crank wearily through without a glimpse of grass-growth. All things wear small green badges of growth, except the grass. The donkey brays more impatiently and plaintively each morning, Fleur whinnies her aristocratic demands and poor shy Syndod stands silent but reproachful, thinner than the others, lower in the pecking order, and I avert my face from all their demanding stares, telling myself that Phyl is absolutely right; without plenty of good hay we simply cannot keep ponies.

Hay, fencing, manageable land and buildings, time and time again they formed themselves into a solid brick wall against which we butted our heads to give us headaches all through those winter months.

Sometimes at bedtime a ghostly whiff of hay would tease my nostrils, as a mirage teases the desert traveller. I took to grabbing the local paper and searching the classified advertisements for hay. I sat alone wondering whether I dare phone Thomas or

Jones or anybody within reasonable distance for the six or ten bales which we reckoned was about our maximum storage load. The forage merchants of course were interested only in orders for lorryloads, and it was so palpably unfair to beg from neighbours who had had the good sense to make good provision for this winter situation. It began to be something of an obsession, this hay, just as something mislaid or lost becomes an obsession and gains an importance in the mind a hundred times greater thereby. I dreamt of haystacks, and Phyl took to adding to her long list of advantages accruing from the electric fencing she was going to buy, some crazy scheme of making our own hay.

"But what with?" I would ask, genuinely puzzled.

She would blather a bit about hayrakes and scythes, whilst I, looking more closely, saw that it was another attack of nostalgia. Grandpa Charlie Kirby had been a dairyman, so there had been an age of sweet innocence when he used to sit his wondrous daughter – I speak of the woman I love – on a sort of throne of straw bales in the long whitewashed cowshed to watch him milk. When she recalled that childhood idyll, the lovely scent of good hay always returned to her nostrils; what she really wanted now was to go back to that old cowshed with the munching cows standing at peace in their stalls on thick beds of yellow straw, and the perfume of hay – and of childhood.

And by God, so did I. Good old Charlie Kirby and his spotless concrete, the good old rattle of his galvanised pails, the singing buckets with the milk spurting, all his endless work. Ah, nostalgia . . . How many times had we not honoured the man and wished him here now. He had done it all from scratch, made it all happen with his own bare hands and a common enough dream. That long white cowshed, every square yard of concrete, mixing by hand and laying it all evening after evening when milking was finished, the mangers and the bottling room, later the automatic drinkers, the lights, everything he had done himself. Twice a day he would fetch, milk and return his cows from the modest little dairy through the back lanes to the commons. Twice a day, every day of every week, he would bottle and deliver his milk, first by pony and trap, later by van, and in the evenings he would load his van with greyhounds and drive up to London from Suffolk to race twice a week, or mend or extend or improve his dairy buildings, or down-calve his cows. He kept his own bull, he maintained his own vans, carted in all his winter keep and carted out all the winter manure, went to market and kept his own accounts, and scrubbed and scrubbed and scrubbed, working as hard as he played every hard hour God sent him, a hard, independent, capable, unimaginative, proud man.

The pile of manure that grew in his yard during the winter months had to be forked into a small cart which a small piebald pony pulled out of the dairy-yard and through the town to a smallholding on the edge of town. Courtship put a manure-fork into my hand, and the reins of that pony, and the courtship suffered not at all because of it, for the little dairy tucked away behind a red-bricked terrace house in a mean lane was as happy a place to work in and to court in as I was ever to know. Ah, yes . . . Come back, Charlie Kirby, and bring your times with you, times when life was settled, and true, and hard.

The memory of those times shamed us all. He and his stock never were caught short of hay as his sheep-loving daughter and son-in-law now were, and he was as meticulously clean and caring in the production of milk as his grandson was fashionably scruffy and uncaring. He would have thrown together an extra building of some sort, of spare timbers and second-hand corrugated sheets, caring not at all what it looked like, and into it he would have carried his two ton of really good meadow hay and sat back with that marvellous feeling of satisfaction that, come what may, one is safe for the winter.

So why did not we do the same?

Dolgwili was an exercise in aesthetic satisfactions as much as anything perhaps, and often to a quite absurd extent. Rather than copy the rural sprawls of corrugated tin additions, the broken-down and rusting ploughs, tractors and harrows, the makeshift fencing and broken gates which were the feature of our countryside, we preferred to go without, even to the extent of having nowhere to store hay. We could not store hay, we could not make hay and we could not buy hay – except by the lorryload. It was unthinkable that we should solve the problem by erecting a proper new haybarn, for a mere handful of sheep, at a cost quite beyond us, yet it was unthinkable too that we could get through a winter without hay. What on earth were we at? We were bashing our brains out and straining our rheumaticky muscles in an endless and foolish struggle. Why did we do it?

We were simply caught up in a flow of nature, a flow of our own nature, a flow of Welsh hill and sheep-nature, a flow that demanded of us all that we could offer – our muscles, our minds, our servitude, our gratitude. We did it all willingly, and were pleased to have been allowed the freedom and circumstances to do it. We wanted only, tomorrow or next year perhaps, to do it better; to get rid of the pasture-disfiguring nettles, to lose no lambs next season, to eradicate the brambles and the bracken, to improve the flock and grass, to solve the hay problem once and for all.

Shut-Up is the pleasant hour. The day is done. The shadows lengthen, the evening comes and our work is done. Then, Lord . . . Then, Lord, you tiredly pull off your boots in the conservatory, peel off your mucky clothes and plonk yourself down in your own old chair in the warm dinner-coming kitchen. If it is your wine week, you sit and sip half a glass of cheap sherry, thinking back over the day. There is nothing more you can do now. The hens are safely shut up, the sheep have been fed and checked, the ponies must please themselves. The shelter and the wood are there if the weather gets filthy. In here the log fire is lit. The old and useless dogs arouse themselves from sleepy comfort to salute your presence with their head on your knee and a tailwag of pleasure. Your shepherdess, you know quite well, is thinking of sheep still, however intent she is on the delicious but frugal meal assembling at her fingertips. Very soon, as Shut-Up glides into Time-Past, when you sit clean at last and glowing before the fire, she will say, "That little gimmer with the black face is a shy little thing." Or, "Have you noticed how thin Nice Fleece has become?" or, "There are two new adverts for hay in the local but they're one twenty-five a bale."

If she does not, you will. Sheep, it was sheep all the way. The sheep of Dolgwili were consuming us, body and soul, night and day.

At Shut-Up, one filthy evening in January, our first Suffolk ewe, a pantechnicon of a sheep, left the end bay of the shelter where we had put her on a clean dry bed of new straw ready for her confinement and walked away unnoticed across the starved and sodden pasture of Sheepdip to drop an ewe lamb in the middle of a desperately cold and torrential sleetstorm and the middle of the meadow. I missed her only when I went to check that she had water, and when I stumbled upon her out there in the drifting darkness, I pointed out to her that it was scarcely a sensible thing to do. She was not put out. The lamb was strong and, despite what we had been led to believe of Suffolks, that they were slow to rise and needed great care at first, she was already on her feet and nosing around for the udder. I lifted the lamb and thus enticed the ewe back to the cosiness of the shelter, saw her comfortable and safe for the night with nuts and water and a hurdle across the entrance. We had lambed our first pedigree Suffolk. It was 25th January.

"Well," my boss said when I reported, "we've just got to get hay from somewhere."

The first three numbers I tried in the following days all brought me busy Welsh farmwives who either could not understand what I was asking or could not make me understand where they lived or how to find it, or had sold out, or did not live close enough to a hard road for our little car safely to get in and out. When therefore I was ultimately answered by a Brummagem accent with words which did nothing to discourage me from fetching a mere Renault-load at a time whenever I wanted, from a farm, moreover, which was only just up behind the Forestry near the village, our village, well then there was little more that life could offer us.

Mr. Taft's farm was lodged, just, on a mountainous slope, so that the vast hay barn stood high above the house's rooftop level. It was still half-full of glorious hay, to me a veritable Klondyke, and I loaded the Renault with solid gold at a mere one pound twenty-five a bale. Four bales went into the car, two on the roof-rack, and while I loaded I gained the impression that the farmer was slightly more than subdued. He seemed content to watch me working and he may just have been ill.

"Oh, you are there," a worried woman suddenly arrived, a distraught woman with much care etched into her face. "You're not helping, are you?" she asked the farmer accusingly.

"No, no," he said quietly, turning away. His jacket, I now saw, hung on him. It had been made for a larger farmer than him.

"Only he must not lift anything," she explained as I handed her the money. "Farmer's lung."

The farm was up for sale. She and their nine-year-old son had been working the farm – there was a herd of milking Friesians – for a year now. They could not carry on. He had always been 'in hay', dealing in it even before he came to farm in Wales. The doctor said he must get right away from hay. The farmer did not know what else he could do, but farm.

"Yes, yes," the woman answered, almost angrily it seemed, as though the burden of the unfairness of life showed in every word. "Yes, come when you like, while we're here . . . Really, we're finished."

The man had disappeared. I thanked her as she hurried away, pulled myself into the Renault and slowly drove up the steep track with my load . . . of one man's meat, another's poison.

21   One more day

It is just one more day.

The lean-to behind the cottage now is blessed with nine whole bales of Mr. Taft's sweetest hay. I fill the hay net with miserly fingers, whilst Phyl measures the morning ration of sheep cobs from the bin into an untidy assortment of half-pails and old drinking bowls. Overhead a new white and tabby kitten named Harlequin stalks bluetits across the corrugations of the transparent roof. Past our muddy wellingtons flows water seeping from the hillside sodden by yesterday's storms. Away by the hedgerows, Fleur whinnies on behalf of the impatiently waiting ponies, starts first the donkey into a bagpipe bray and then, further off, the elitist Suffolk ewes and hoi polloi Speckles into plaintive demand.

As we trudge through the waking garden, the animals beyond grow rowdier. Hyacinth and heather, primrose and forsythia momentarily improve my mood, then I am beguiling the ear-pricked ponies towards the orchard with hay net, nuts and calls to allow Phyl to enter Sheepdip and the wheeling scrum of hungry ewes.

Wary of hooves and jerked heads, I spill small piles of pony nuts strategically below the black apple boughs, hang the hay net, slither through mud to safety between the jealous ponies, quickly down to feed the paddocked Suffolks whilst Phyl copes with Speckles.

"Weighday, isn't it?" I call as I pass, but she is heavily engaged.

The three expensive Suffolks have produced four early lambs of disappointing growth. They all charge at me ravenously, prod my thighs with cadging hooves and gorge the nuts as I sprinkle them into their trough. As I lunge cleverly to catch a lamb for weighing, my wellingtons skid from under me, throwing me down swearing on hands and knees in cold mud, whilst the cartoon-lamb stands laughing its black head off. Not until Phyl comes can we entice the Suffolks back into the shelter to weigh the four lambs.

She holds the lamb, I wrap a gaudy sheet about its middle and we argue its perfect hooves through four cut holes and grab at the strings from which now it will hang suspended from the old brass scales. It bounces stiff-leggedly in mid-air whilst I record its daily gain and Phyl comforts it. Disappointment flows as we discuss why the lambs aren't doing better.

As we remove the holding hurdle to release them, the massive Suffolks start a crazy rush and somehow Suffolk Number 8 plunges her great body between Phyl's legs and in a whirl and a cry the best wife I ever had goes off saddled to a runaway ewe, rodeo-style, until she slips off backwards on to her bum in muck, mud and indignity. I swear on her behalf, pull her up, and somehow we shepherd them through on to the lawn itself, their special domain of reserved grass enclosed in our new electrified fence.

"You all right?"

She is limping just like Charlie Kirby and has mud on her cheek.

"I shall have to go and change my trousers. I'm wet through."

"I'll get Syndod into the stable and give her extra grub. I'm ashamed to see her so thin. You stay in and get coffee."

"We're three short on Speckles. Gentle, Daisy and Darkie. Can you have a look round?"

"Okay."

"Only, Gentle has that bad leg and all three are due."

"Right."

I ski with head down through the mud to open up the stable. The magnificent Light Sussex cockerel leads out his ten hens from their scratching of the brown bracken litter, blinking at sudden sunlight, bopping to poop in gorgeous black and white. Already Syndod had dislodged herself from the bossy gang and comes to me begging for alms. I lodge her in the stable and fetch her a good bowl of nuts. The line of her backbone is too distinct and I shut her in to feed, whilst the black and white chickens fussily march towards drier ground. Up in the orchard, as I set out to search, Fleur ill-temperedly threatens any other pony who nibbles at the swinging hay net. All across Sheepdip the Speckleface ewes have fanned out, heads down at short grass.

Huge cumuli, swollen with dark rain and hail, drift across the clear sky as regularly as the silent transatlantic aircraft which scratch white trails across the same blue sky. As I climb towards the Steep, the claustrophobia of the valley eases, green planes of earth tilt gradually and one becomes more aware of wind, of the balancing buzzard, of the stretching plateau of hill pastures.

The ram, Golden Boy, has developed the charming habit of attending his wives in their confinement, and now he follows me a little hesitantly up the hill in search, perhaps, of Gentle, Daisy and Darkie. Four magpies, uniformly black and white as penguins, bounce along the grass down on Sheepdip, whilst I search each hidy-hole among gorse or dead bracken as I plod upwards. High above me, near the wood, old Darkie appears out of scrubby patch; I tick her off my mind.

I climb the slippery and jagged outcrop of slaty rock off the track on to the grass at the bottom of the Steep and, alongside a run of blackthorn hedge, there is Gentle. She is on her side, straining. No lamb shows. She is comfortably enough placed, sheltered, on a slight slope, yet already she disturbs me. As she lifts her head and stiffens her thin legs to strain again, she seems tired. I find myself shaking my head, muttering. I bop down and watch.

We have never yet lambed an ewe without help. We have said repeatedly that we must. In twenty minutes or more, Gentle has made no progress. Time begins to crowd me and I argue back that nature must have its way, interference could be wrong. She makes a token show of fright when I close in to examine her, but already she seems too weak to flee from me. I lift her messy tail. Within the cervix two small wet feet show. I am not sure what must be done or when; nothing, anyway, without a bucket of warm water, soapflakes and towels.

All the way back up the hill after coffee there is no confidence.

"Are *you* going to do it?" I ask.

"No. You."

Why's that? I would have thought she would have been better at it. Move your hand as carefully as if you were fingering

172

sheer silk stockings. Allow the ewe at least four hours . . .

"But we don't know how long she's been in labour, do we?"

I remember with a rush of anger all those slap-happy shepherds on television, casually pushing into ewes and pulling out lambs as easy as picking apples.

"I wonder if this is anything to do with her limp?"

"Poor old thing."

Do not go gentle into that good night . . . She is a favourite, this one. She looks more exhausted now. The two feet show.

We make camp, spreading coloured towels on bushes, tipping soapflakes luxuriously into the bucket of warm water, trying to remember what the book says. I take off my coat, push up my sleeves, soaping up. Shadows of cloud chase across the hillside. Golden Boy appears a way off, pacing corridors.

My soapy hand has no confidence. It splashes and cups warm soapsuds again and again on to the cervix, tenderly explores with slippery touch to recognise reality from black and white diagrams on the page of my brain.

"It's all so tight," I have to complain. "There is no room. One finger is all I can get in beside the legs."

"Can't you push the legs back in?" she quietly teaches me from her kneeling place by Gentle's head. As though they were hers, my fingers probe, push, ease, but all the time I feel that this is wrong, abnormal. She has been like this for hours now.

"It's all so tight," I issue a duplicate bulletin, delaying, delaying. "Isn't there a condition called ringwomb, I seem to remember?"

Cold blew into us and the day passed us by. The world had dwindled to a single aged ewe and a lamb struggling to be born. Intent and lost, we simply did not know what to do in safety. Still the ewe strained, still my fingers moved tentatively. And then, quicker than suddenly, the head was out. My encouraged hand grew more demanding and the legs were out too.

"Good, good. Ooh, I shall have to stand up for a minute." She staggered a little on that Charlie Kirby leg and we smile at each other. In a minute she would say the lamb was dead anyway . . .

"Ough, I'm getting cold. Throw me that towel, will you?" and I move about briskly in the sun and wind, like a weightlifter 'psyching' himself up. "How's your poor knee?"

Never mind her poor knee, what about the lamb? Pull only when the ewe strains, remember.

"She doesn't strain any more anyway, poor old thing."

It is like pulling on a live fish; there is nothing to hold, nothing but slipperiness. And it may be dead. Nevertheless, I kneel and pull, and pull, and pull, measuredly, spasmodically, an elderly man, an aging woman and an ancient ewe in ritual stance, on bended knees in reverence of new life.

The lamb slides out suddenly. Shapelessly it flops on the grass. Well done, somebody says far off. Is it alive? There is no movement.

"Shall I swing it round?"

That I can distinctly remember. The book had saved more lambs by swinging round than in any other way.

I am damned if I am going to let it not live now. My fingers are eager, aggressive even, as they push into the clenched mouth to clear away the mucous from between sharp teeth. I lean down and breathe life into it and I'm shouting in its black wet ear, "Come on, you in there, come out and live. Live. Live, blast you, live!"

And it lives. Dammit, it lives. You great big beautiful God, you; it lives. Look, it's shaking its head . . . And its leg moved then. It is alive.

The slimy lamb confirms its existence with groping movement, anxious for the reassurance of the mother's licking tongue, which does not come. Gentle is spent. She is panting, lies her head back on the grass and is quite without interest in her lamb. Each time we lift the lamb to her head she ignores it.

Whilst we towel the yellowed lamb dry, Gentle whimpers quietly and starts to strain again.

"She's got another . . . "

Her second lamb, alas, requires no midwife, no nursing; life for him is already over. The waste, the waste, we sigh, talking to Gentle, worrying, worrying.

Much later, Gentle still cannot stand. The blanketed lamb, blaring for sustenance, is held by Phyl against the washed udder where its black nose explores all the wrong places for the first vital teatful of colostrum.

"Right. I'll phone the vet. You okay?"

She nods and I step outside the intensive care unit back into the uncaring world of grass, hawthorn, distant cars, down the track to find a mortuary for the dead lamb, to phone the vet.

Through the window, beyond the reddish haze of burgeoning prunus, a Suffolk lamb is challenging our electric fence to nibble my golden variegated holly. Suddenly the pulse shocks his nose and he shoots backwards and goes off sulking. This morning his docked tail fell off. If I had more strength of will and my fingers could get the knack of positioning the tight little red rubber rings over his testicles I would castrate him, for already it is obvious that he is only half the hoped-for size.

The vet arrives within the hour, all brisk and smiling confidence, selects medicines from the boot of his car, demands information, throws on his white coat and strides ahead of me up the track.

He releases afterbirth, gives three different injections, diagnoses post-parturitional paralysis, deals out advice and Pendipan Forte to be injected over the next two days. The limp had probably been caused by a lamb pressing on a nerve, and yes, it well might have been ringwomb.

I envy the man his confidence. He cleans up, dashes away with our gratitude and we leave Gentle to rest and recover.

"You go and get something ready to eat, I'll look for Daisy."

My back aches. I stroll towards the standard lambing places. The undisciplined hedges, mostly of thorn with young ash but with two very old laburnums, are thin and of use along this track only as windbreaks.

Perhaps I am developing a seventh sense, but once again something leads me correctly in my search. Wind-direction, or storminess, or the season itself, and habit, all help. Now I have come directly upon Daisy.

She is grazing hungrily, quite separate from the flock. Almost certainly she will lamb here sometime today. Because I do not wish her to come to me for nuts, I dissolve into the hedge and freeze, watching the ewe. Even as I do so, she becomes alert, stops eating and stares at something along to my left. A fox enters left of stage. Ewe and vixen – is it? – stare bright-eyed at each other across twenty yards of cold March air; I stare at both. We are all quite still. The fox's mouth is open, laughing perhaps, and one paw is held off the ground. Challenge is in the air. It is like the stopping of a film. Action rolls again. The fox turns back towards Thomas', trots away and then breaks into a gallop, the ewe resumes grazing, and doubtless I smile a little as I exit right.

Good . . .

Gentle has not moved. I am very hungry.

The simple foods are joy. The stuff they call muesli, with cut grapes, honey and apple lumps, a single biscuit loaded with cheddar cheese, a steaming cup of coffee, and all the poor World-at-One suffering its flaming head off, and poor old Dan Archer beggared of sheep . . .

" 'Ow's yer poor leg?" I ask again in exaggerated jokery as we pause for rest from pushing our little two-wheeled trailer up the slopes of Sheepdip towards Gentle. My wife confesses that she has also suffered electric shock at the hands of Suffolks this day. The silly Suffolk lamb had this time entirely entangled himself in the electric fence in further attempt to eat my young holly, and Phyl, instinctively fearing more for the welfare of both lamb and fence than for my holly, had rushed down to disentangle them, without first switching off the current. It is a very disconcerting shock the fence gives.

The ewe lamb is wrapped up and contained in a Beanz box in our trailer-perambulator, yelling its head off again for maa, maa, maa. We zigzag up the slopes awkwardly; all planes, holds and angles are wrong, everything difficult except the surface we traverse. All the grass has been eaten down and we walk a smooth carpet of the flat rosettes of meadow daisy.

Up there we rest again, suffering a hailstorm. High above, the jets still fly off to kingdom-come.

The ewe still cannot walk, but soon all four of us stagger down the slope a little at a time towards the parked trailer, tied to each other in helplessness like prisoners. At the trailer, suitably tipped, Gentle almost climbs in like a little old lady and I half-expect her to throw a feather boa round her thin throat as we settle back for the downhill slalom to the stable.

We have gained a pause. Gentle and lamb are safe for now. From here we can assess what remains of our day, all our routine feeding, shutting up and resting, plus the two ewes yet to lamb.

In the early evening I find Daisy with a strong ewe lamb. It is sucking, there is no sign of fox. I collect the lamb and with her entice Daisy down to be shut up with Gentle. But I find that Gentle has started straining again and seems to have very little milk. I have never liked drying lambs, of coming between mother and child; I fear this will become a bottle-fed lamb. When I examine Gentle I find the neck of the uterus is showing. She strains again. Damnation!

The vet receives my late call with patience, and I receive his advice with apology, for what sort of sheepmaster is it that does not know of prolapse and its treatment?

So we abandon the evening's comfort, throw back on our old togs and drag ourselves back to the stable with more towels, salt and warm water. Gentle, despite illness and age, fights us strongly all the way as we raise high her hard legs and hold them there whilst Phyl bathes the neck of the womb with the warm salt solution. The misplaced organ slips back surprisingly quickly. It is just as well, for I cannot hold her up any longer.

My head throbs with the effort. My bruised and electrified wife consoles the ewe, and I know she will be aching as well in sympathy with the animal. By now the lamb is sucking at everything within reach, so we decide that Gentle and lamb will be better off parted. Back in the kitchen we shed soiled clothes and Phyl prepares a bottle. The lamb, once it has learnt, sucks in nearly a quarter-pint of warm, glucosed cow's milk.

An hour later I return it to Gentle for the night. Both of them still are cause for concern.

I check the Suffolks, and go back again for a last check on Gentle. All are comfortable. The day is done.

The stars are out, the wind is keen. Near the house the night-scent of hyacinths under the rosebushes halts me for a moment, then I drag into the warm and welcoming home, and sink into an armchairful of ease. The marvellous 'Petersfinger Cuckoos' has returned; it is beautiful beyond words.

22   A sort of renaissance

Before ever Dolgwili coyly offered herself to us, we had bought, or considered buying, sufficient other properties to know that a most useful corrective or supplement to those often irresistible details supplied by the house agent are the more down-to-earth opinions let fall by the local publican, parson or plumber. In the case of Dolgwili, most fortunately, we did better even than that, for we had the opinions of Mr. Powell, a local gentleman with whom we stayed and who talked his way into our history with his phrase, "It's right in the eye of the sun."

"Has possibilities, but what about access?" we had marked our agent's particulars when first we had viewed. Even the agent, when he accompanied us on our third viewing of the property, had been unable to conceal his qualms about access. He had sat in a rocking chair in the sunshine out front of the house which was to be sold furnished, itself an indication of how bothersome access was, and as I approached him after a long prowl around the place, I fancied that he was purring at the prospect, scenic and financial, of that moment.

"Um," I said, allowing him time to be a little more alert before I fired my loaded question, "yes. It's . . . er . . . Well, we can see all sorts of potential to the place. We do like it, but . . ."

Naturally I paused and looked keenly all about me to emphasise my point.

"But how exactly do we get up here with the furniture van?"

"Do you know," this most pleasant gentleman smiled and now admitted, "I sat here just wondering the same thing myself."

He stood up and joined me in my keen looking about. Down the slope, a quarter mile as the jackdaws flew and all among the trees, the tangled skein of road, railway and river strung out along the valley, and always the road was on the further side. He pursed his lips and shook his head, turned round and looked up and over the roof to what later we were to name the Steep, and, apparently quite seriously, he suggested that perhaps the furniture could be lowered down that slope to the house.

We received the suggestion with silent incredulity but with politeness. We were, after all, in a new country. It was quite extraordinary how uncertain we felt ourselves here. Perhaps it was customary hereabouts to lower valuable grandfather clocks by a sort of removals cable-car down several hundred yards of rough terrain. Perhaps they had even developed a special technique to swing fine furniture clear of the gorse and hawthorn that abounded there. We should, not, however, ourselves favour the idea, for we knew something that evidently the agent did not and, who knows, it could be worth a reduced bid to us.

There was a ford across the river, our Mr. Powell had said, and indeed there was. We had already inspected it – it was very shallow – and we had walked the track that led from the ford, along by the splendid river, up and over the railway line and so to Dolgwili. It was not exactly ideal for the transportation of grandfather clocks, nevertheless we did know for sure that there was vehicular access of a kind and that certainly we could get our furniture up to the house.

As a bargaining ploy, however, our knowledge and the agent's ignorance of the ford was of little value. Its significance was almost totally swamped by the much weightier fact that the railway itself was closing down completely and for ever next month. The agent knew this, Mr. Powell knew this and no doubt publican, parson and plumber knew it. The only present traffic on the line was the famous milk train which trundled and squeaked up the valley in the morning and back in the after-noon, a long slow affair of coal trucks and milk tankers, un-economic and doomed. Next month it would cease. After that, obviously it would be a mere matter of months before we could buy that short stretch of railway with its vital bridge high over the river and have the sort of civilised access to which such a place as Dolgwili had every right. Everybody knew this; every-body told us so.

*   *   *

Well, four years later . . .

God, how we grew to hate that damned stretch of railway. Day after day, week after frustrating week, those cursed railway lines lay immutably there, so damned parallel, so damned straight and impregnably solid, useless, obsolete, rusting, and immovable. Dead, and so obviously dead, now all sorts of saints came forward to resurrect it. The weekly paper had us on the rack with its alternating news. It was to be taken up, it was not to be taken up. BR had said that this was definitely the last extension of time and next month it said it again. It was to be taken up, it was not to be taken up, month after month, while we aged and muttered and staggered up and down from house to road carrying everything we needed to live along the two hundred and fifty yards of pedestrian right-of-way over the railway bridge above the moating river. All through those four years of slumbering negotiations to buy or lease or borrow those forty vital yards of railway track, which contained the marvellous fortress-like bridge of stone, in all that time no vehicular traffic, other than a wheelbarrow, carried our groceries, our coal, our animal feeds, much of our building material and all the daily paraphernalia from road to home. Our furniture, all the heaviest materials, and once our coal, came across the ford by tractor and trailer, and that only at picked times of low water.

At last some courageous or desperate man in a distant office lost patience perhaps, sent in an incredibly efficient gang on a special train to whip out rails, cups and sleepers in what seemed comparatively only a matter of minutes after all the long months of waiting, and shortly afterwards we were figuratively pouring champagne on our famous bridge and blessing all who would cross by her.

A sort of renaissance followed. In a kind of reversal of the opening up of the American Mid-West by the railroad, Dolgwili now was opened up by the removal of the railway. A steady traffic of regular Renault-loads of Mr. Taft's hay bales, seven per load, of weekly groceries and animal feed, even of a real lorry loaded with anthracite, followed and transformed our lives. We began to walk a bit more upright again, and to realise all the things which now we could do that had previously been impossible.

It was a heady time. Soon a young man on the largest tractor

I had ever seen was hauling a trailer across the bridge repeatedly to spread loads of rock phosphate and lime in swirling
white clouds of dust all across Sheepdip and Garden Paddock,
probably the first time in thirty-forty years that anything had
been added to the farm. The brave little ride-on mower with its
garden trailer, our only machinery, now could get across the
railway track and down to the river for sand, stone or logs and
really, we felt in that 'glorious dawn' there was nothing now
that we could not accomplish.

Mr. Thomas came down and trimmed all the hedges for us,
we made ready with awful chemicals to destroy the thistles and
nettles that all our laborious work with scythe and spud had
over the years failed to do, and we were never more definite that
this year we would eradicate once and for all the ever-spreading
bracken. We should castrate and weigh, inoculate and regularly treat against foot-rot, we should eartag the flock so that we
could instantly recognise which lamb belonged to which ewe
and generally we would build up a wonderful flock of superb-
looking and thriving animals. Above all we would improve our
grass – and the instrument, already in our hands and proven –
was the marvellous electric netting. With it we could erect
paddocks of considered size which would contain the flock with
no worry of them breaking through. Whilst the flock grazed one
paddock we could prepare a new one in advance. Management
all round would be easier and more efficient. By God, what a
time we were going to have . . .

There was little doubt in our minds that we were becoming
ourselves what we had always thought of as 'proper farmers',
and thus, over-confident and temporarily blind to the fact that
Dolgwili could never be a 'proper farm', we began to search ever
wider horizons in our quest for better sheep-farming.

Even before our obsessive need for hay had sent me to regular
reading of the local paper's advertisements, I had found the
small, classified adverts by far the most interesting parts of the
whole paper. They needed to be read in searching detail, for all
sorts of apparent riches tended to get their mention among
dross. Pedigree Suffolk tups with almost royal prefixes or
Registered British Toggenburg in-kid nannies (we were getting
interested in goats now) might easily find themselves penned
and on offer alongside 'Ferguson tractor for spares', 'Baby Cot

(needs new engine)', 'Colour TV £10 or would exchange for Council House', mink coats, quantities of wire, swimming pools, Jack Russell puppies, Welsh dressers and tiled fireplaces.

"What on earth are 'One Scrown Potatoes'?" I once had to demand of Phyl, and always these small-ads bulged with the most illuminating comments on the daily life of our rural community. Having graduated, however, in sheep-farming, as we thought, we now tended to graduate from small-ads to Clear-Out Farm Sale announcements. We took avidly to searching them for drenching guns, hurdles, feeding troughs, hayracks and any sheep paraphernalia that might assist, at reasonable prices, our ever forward and upward march. Sam-Ram's long trek from Dolgwili up to the dip near the village and the stress it resulted in had tentatively put a sheepdip of our own on to our shopping-list of desirable agricultural requisites, and so a dip was always something we searched the sale announcements for. A sort of husband versus wife competition, of no real significance, set in, to find a second-hand sheepdip among those week-after-week announcements that always started 'Having sold the farm . . . ', or 'Due to retirement, Mr. Handel Rees' or 'Due to change of farm policy'.

Sheepdips, of course, are for the most part built-in and immovable concrete-and-iron constructions like the admirable one we had so coveted at the farm demonstration last year. The best sheep-farmers never had stopped dipping their sheep, even when it had not been compulsory by law, looking upon it as good husbandry, and these farmers obviously would have easily-managed and correct installations. Since the reintroduction of regulations on compulsory dipping, however, a new dip, pre-formed and portable, a sort of narrow, deep bath with steps, had come on to the market, ideal for small flocks, it seemed to us. This, second-hand, was what we competitively sought, and this was what never once came on offer, until now.

"I've found a dip," my wife crowed. "There's a grass harrow in the same sale."

Well . . .

We did like Clear-Out Sales. Quite apart from that mad hope of picking up a gold-nugget at such a place, they were revelations to us of how other people coped, or did not, a sort of dissection of that farmer's work-style and even life-style, a lifting of the stone, as well as a criterion by which to judge one's own work and style. Still, this sale was quite a long way off, by my own shrunken travel standards. And really it was quite ridiculous for people like us to consider installing our own dip. And there was so much more vital work to be done here, and well, really I had very little appetite for the idea. Nevertheless, being her father's daughter, Phyl was little inclined to relinquish a dip which had been so difficult to find for such flimsy reasons now, so she determinedly filled thermos flasks with milky coffee, cut piles of cheese sandwiches and flaunted before me a whole packet of my favourite chocolate-digestive biscuits, happy in the knowledge that she was about to set the seal on our farming accomplishments at a mere bargain price.

It was spring. It was sunny. It was nice to be away from the toils of Dolgwili, a sort of holiday. We travelled easily and with a pleasant sense of anticipation. The main roads became by-roads, and then lanes. The sale was even well sign-posted; all the auguries were fair.

As we turned into ever-diminishing lanes towards the farm, the unavoidable impression was that half Wales was arriving for what was, after all, only the viewing day. Half Wales, too, had been examining farm adverts for mention of a sheepdip. Cars arrived from all directions to crawl to the farm, there to join half-a-hundred already untidily abandoned alongside the road, on verges, in gateways, in and about the farm itself. There were even a few in the official carpark, a higher meadow with a narrow gateway near which a few straw bales had been littered to neutralise mud. From each vehicle people emerged, it seemed to me, hurriedly or anxiously, so that I began to fear that we had things wrong, that this was indeed the day of the sale and that the sheepdip was already on offer. There was some extra charge in the keen air, an excitement or curiosity, as we locked the Renault and joined the strollers and viewers. Cars arrived with a strangely strong impulse to park anywhere, and the only official I could see, wearing a white overall and dispensing catalogues, already seemed bemused and helpless within an invasion of cars, Land-Rovers, vans and even lorries. The small area near where he stood was choc-a-bloc with men and women milling about, clutching their large white sheets of catalogue, crossing the lane, peering at the Land-Rover there for sale, the chainsaw, the car, the tractor, looking about them like strangers at a loss.

"Surely they can't all be here for the dip," I said again, repeating my joke half-fearfully, and starting our tour of inspection.

Certainly they could not all be here for the dip. It was merely a galvanised tank, large enough to take one of our huge Suffolks, say, and portable. It was indubitably a sheepdip; it was, indubitably, not the honey-pot that could possibly have attracted this swarm of interest. We kicked it once or twice, and I bent about it in postures of examination to show bystanders that this deerstalker and these green wellingtons were not merely rural showpieces but that they denoted a real sheepman who understood dips like this one. Privately, however, we muttered to each other that really it would hardly be of more use to us than the rusty old collier's cart which already stood semi-derelict in that corner of Dolgwili, and so, crossing it off our catalogue, we moved on to the grass harrow.

At that period of glorious energy and ambition in our sheep-life, few other things seemed so desirable and prestigious to me as a grass harrow. I see now how wildly ambitious I was, and indeed the harrow itself, as it lay there by the haybarn by itself, almost declared me so.

What I sought was a heavy, spiky, towable implement that would tear and comb out the accumulation of matted dead grass of wintertime and thus allow the new shooting grass of spring to grow free in space and air and light. The improvement of the pasture was high in our priorities. A grass harrow had become as obsessively desirable during our advert-searching days as had a sheepdip.

Lot 76, grass harrow, would have done my job perfectly. I leaned down and tried to lift a part of it. I kicked it. One has to kick these things at sales. It denotes a familiarity, a knowledge. "By God," those nearby me must have thought admiringly, "by God, I bet he's known some harrows in his time."

"Will it do?" my marvellous woman whispered in my ear.

I motioned her away from it and we stood conspiratorially in consultation lest anyone should hear what I had to say.

"If, somehow, we could snip off about one square yard of it. And if, somehow, we could get about four good strong men to lift that piece into the back of the Renault. And if the front wheels stayed on the ground, well, then, yes, yes it would do."

"You mean it's heavy?"

I nodded and looked sad. It was always sad when you came to a sale and did not buy anything. But I shuddered to think of the strength of Edwardian men and their horses.

That left only the silver. We would look at the silver in the farmhouse after lunch. Not that disappointed, we wandered back by the rather cosy farmyard, unhurriedly gazing at chicken coops, baled hay, a turnip-chopper, a horse plough, odds and agricultural ends, drifting towards the car. In the orchard snowdrops were still in bloom. Everywhere the crowds thronged. A small greenhouse entombed the black skeletons of last summer's tomatoes, garotted by twine. It had been roughly repaired in one corner with boards. Other shrivelled plants, still tied to grey canes, had been grown in old feed troughs. A geranium still clung to life. When we peered into an outhouse, strings of onions hung there, golden-ripe and impressively neat in their stringing. They too, we looked up, were catalogued and no sooner had we stopped peering than other peerers took our place.

There was something most pleasant, it seemed to us, in the setting of the farm. It had the Presceli Mountains as a backcloth and fine views all round. When we strolled into the pastures around the farmyard we began to drool how fine and high the grass was and how kempt the hedgerows. Coo, we cried to each other, overcome with envy again, how easy it would be to farm here! Forty or sixty acres or whatever it was, how well and how properly we would farm then.

The little red car was a haven after all the sale business. We sat and munched cheese-and-pickle sandwiches and sipped at the steaming coffee, gazing out at the enviable farm. The distant hills and trees slumbered, tired of everything. Outside the air was keen, here inside all was cosiness.

"Is it worth staying?" I asked.

"Ooh, yes. Now that we are here."

"So many people," I grumbled.

"Yes. I wonder why? Those onions . . . "

Yes. Something seemed unusual.

A substantial queue had formed to get inside the farmhouse.

The silver tea-service was catalogued as 1939, a few antiques sounded a bit interesting, but nothing catalogued explained the still-swelling crowd of viewers. As we drifted towards the queue, an outside stone-stairway to the barn invited me to ascend it. An open door at the top led into an almost empty, dirty room. In it one man stood, examining with amused interest an object he had just picked up from the dusty windowsill.

"It's years since I saw one like this," he said, seeing our curiosity, "Do you know what it is?"

It was of yellowed horn. I turned it in my fingers. The tip of the horn, a cow's I guessed, had been sawn off.

"I've never seen one like this, but if I had to guess I should say it was used for drenching cattle."

"Aye, aye, that's it. Many's the time I've used one as a lad."

Many's the time I had sold one as a lad, not of horn, mine, but of tin, with a wooden gag also. It was of no importance, the horn, except that it had brought us into conversation for the first time with somebody at this somewhat unusual sale.

"Why on earth are there such crowds here?" I asked the stranger in due course.

The farm had been owned and run by an elderly brother and sister who had lived here all their lives, he told us. About a fortnight before Christmas he had murdered her. Then, after trying to burn the house down, he had turned the gun on himself. One story said they had quarrelled about money, that she held the purse-strings and that eight hundred pounds in cash had been found in her purse and considerable other sums of cash about the house when the police investigated the tragedy.

"Aye," the man said, broodingly, looking down from that upstairs window to the crowds of people gawping around below. "Aye . . . "

We became gawpers ourselves. We stood for a minute or two with our informant, brooding with him, and then went down those stone steps to join the queue to get into that house of tragedy. Inside, the small rooms were made atrociously smaller by the shoulder-to-shoulder crowds who jostled and shuffled in disorderly procession up and down the narrow staircases and through the dark and dirty rooms. The wallpaper, the doors, the furniture, everything was coated in a dark film of paraffin fume, which left a pervading smell. Still on the table in the place which it had occupied since wartime evidently the silver tray, on which a rather attractive silver tea-service showed the brightness of silver only when some bolder soul picked up the teapot or basin or jug for cursory examination. A good-looking old cheese-dish in the form of a bull's head showed its cream colour only where somebody's fingernail had scraped the black film. The bathroom caused one to shiver on sight of it, so cold and stark did it seem from the small doorway. It was long and low, an attic just under the slates, with the old-fashioned white bath standing in the centre of the bare boards and the water-pipes groping barely, like arms, across space to reach it. A single bare electric light bulb hung there, dead, illuminating now only the spartan existence of those who had shivered here once. In one bedroom an iron bedstead stood graceless and unclothed, bearing piles of folded blankets on its lap, all, everything cata-logued in this ultimate of clear-out sales. We shuffled along half-ashamedly, shouldering our way in silence through that place of darkness among a chaos of men and women come to gawp at the final mystery of life, death . . . and murder.

"Oh dear," perhaps one of us murmured as we wandered back towards the car, past a thousand budded daffodils in the grass near the house, our minds possessed by mystery. Why? one asks all the time.

In the narrow lane the traffic chaos still increased; cars tried to leave, cars tried to arrive. It had been the same inside the house; nobody directed, nobody controlled. People swarmed, their workaday selves discarded, their minds taken over by some elemental need to come to this place of tragedy.

"Well? Do you want to come to tomorrow's sale?" I asked, as eventually we disentangled ourselves from the running maul of cars in the lane and drove out into unsullied country.

"No."

23 They wore our lives down to the bone

Before the successive onslaughts not only of sheep but of all their battalions of allies – moles in the lawn, rabbits in abundance, slugs, caterpillars, mice, frost-pocket and cold clay – I came to realise that I had nowhere to go but backwards. When at last I decided with sadness that it simply was not worth the fight to grow vegetables, I knew in that same moment that my old and spasmodic affaire with the minx called self-sufficiency was over.

I turned my back on her with real and almost guilty regret. It was, I called back to her, the sheep, you see. I was not my own man any more. Time, I shouted a little desperately, the sheep take all my time.

So I abandoned Malling Promise and Glen Clova and allowed my hard-won vegetable plot to desert to the all-conquering grass once more. Constance Spry, Mrs. Gomer Waterer, Contesse Bouchard and the rest of the aristos I loaded on my tumbrel like refugees and together we retreated to our stronghold nearer the house. There we drew a last line of defence beyond which I swore no sheep would ever pass. That line, undefended and imaginary all through summer and part of autumn-winter, was marked anew come spring by the yellow palisade of electric netting, aesthetically displeasing but fearfully comforting. It ran from the Sheepdip side right across the garden to the Lower Rough side, near the top of the slope, high-waistedly, Empire-lined, just down from the rockery screes and the foliage border. This was the heartland, I bravely proclaimed. Below the line, There Be Sheep and even dragons, but this, this is my kingdom. 'They shall not pass', I engraved on our new garden-seat with my penknife.

So by summer, with all sheep contentedly banished from this

188

entire garden area, I ruled. The foliage border became all my joy.

It had started with a single beautiful shrub, ugly named as cottinus coggygria foliis purpureis, a cube of purple foliage which positively glowed in sunlight. Darkly lush, it was the equivalent of the first statement in bold paint upon virgin paper, and having made the statement, it was simple to add nearby the contrast of silver-white cineraria maritima. Not only did silver contrast with purple, but the finely dissected leaves of one contrasted with the simple oval leaves of the other, and then having started this plant-painting, old nature seemed to join in the fun and threw a casual paint-pot or two itself. The shiny green and solid leaves of bergenia arrived close by unplanned, lower to the ground and indicating other dimensions. Soon the mind's eye opens perceptively to see the need for something pencil-slim and tall, golden perhaps, to grow here, and something prostrate or rotund to grow there. Needled yucca, luxuriant pampas grass, variegated hollies . . . Colours, shapes, textures.

At times the whole thing, the concept itself and even the messing and grubbing about down there on hands and knees in the earth itself, the whole thing grew in joy to become an intoxication. A lily growing to white perfection against dark foliage, generously perfumed and spiring; the shapeliness of juniper; the long-loved Lady Sylvia rose modestly thrusting just a few perfect blooms through the purple cottinus . . . Instead of the pursuit of fame down the corridors of court, how much more pleasing it was to occupy the mind and fingers with this quiet creation of private beauty, to bend the padded knee in unconscious reverence of great mother earth, to peer through spectacles which always drifted or hung at wrong angles on a nose which always needed wiping but never was, so intent was the old fool on his care for plants and his need for beauty. Sometimes the heart seemed to be quietly bursting with gratitude for all this fulfilment and with all the promise stretching away before me.

Ah, life, life . . . Even now, growing dim and bald, socially redundant and low in the pecking order, still I was besieged by the demands of life, possessed by the need to procreate. The purpose of life is life. Life everlasting; not my life, not your life, but life's life. Continuation, evolution, change, wherever you looked, whatever you considered, the reason for every flower's shape, every creature's behaviour, all creation had but one motivation, one great purpose; the perpetuation of life.

Yet, even in summer, my gardening days came ever more seldom.

Hardly such a day passed but that the sheep called me from my knees with their trespassing, or their sufferings, or their transport or their moving or their feeding. The time of no-grass passed to the time of too much grass, the time of brown dead bracken to that of green live bracken, and although the summer sheep could now contentedly graze, both long grass and tall bracken meant that winter sheep would graze even less contentedly if both were not cut and cut regularly, and now, before they were too long to cut at all. Garden Paddock in particular grew amazingly as the eye of the sun lit on the slope lately enriched by rock phosphate with lime, and more recently by the rich guano of sheep droppings. Round and round rode Rosinante with myself playing Quixote, a weak-headed ancient suffering the delusion of being a sheep-farmer, and thinking to make, of all things, silage of this surplus grass. The lawn itself needed cutting three times a week in May, an easy and even pleasant job, but one to be followed for poor Quixote by the hard and long jobs of raking and gathering. It was almost impossible to see all that lovely grass simply going to compost when one remembered back to the awful winter shortages, and so, without confidence, we let it dry off in the sun, heaped it awhile, and when it seemed rather nicely limp we stuffed it into large polythene sacks and stored it round the back of the house.

No sooner were lawn and paddock smooth and seemly for a day or two, than poor little Rosinante was being driven up and all over Sheepdip by her Quixote who now was stricken with the notion of eradicating bracken. Other giants – Nettle, Thistle, Gorse – were on his list too, but bracken, that was the Big One.

"Cut it at the early stage, just as the first fronds uncurl," the specialist had advised us. It was like the mosquito-cure directions on the packet in the Klondike in '98, 'Place the mosquito on one piece of wood, kill it with the other.'

Somebody should do something about bracken, other than

trying to kill it. If bracken could be used, what a crop, what riches we would all reap. Dolgwili and the rest of Wales grows bracken as tall as a horse and almost as strong, on every hill.

So, husband on mower, wife on scythe, all through the summer weeks we worked, all for the sake of sheep. We drove and cut in swathes and when you turned round to see where you had been it had started to grow again. It grew before your very eyes. "Just at the early stage as the first fronds uncurl," is all the time. Day after hot day we continued, as much as we could for as long as we could, whilst her fingers and my bottom blistered, whilst muscles ached and numbness set in. Often the slopes of Sheep-dip were too much for the gallant little mower and for his rider. Then, instead of cutting up and down we would have to change to along-the-slope cutting, with Quixote leaning out like a side-car rider or trotting alongside the reverse way. Occasionally the side-car act grew more hair-raising when my hip was so low as to disengage the gear-lever, for then the mower had no brakes. Sheepdip then was treated to a Quixote-charge down-hill, brakeless and with little steering control but with total belief in the efficacy of prayer to avoid all the little anthills and sudden changes of contours which could have turned us over for ever and ever.

Wherever the face of the pasture was disfigured by these old anthills, like warts eighteen inches high, grassed over and de-serted by their makers, Rosinante was disinclined to take me. Here it was scythe work, woman's work, never done. Even on the smooth parts, where the mower's fiercely rotating blade spewed out a chopped brown mess, there were always parts where the scythe had to clean up and always this meant conse-quent raking, piling and carting. It would make, we used to assure each other, very good litter once it was dry, and so we stacked it out on the sunny hillside with the feeling, such is the limitation of the climatic imagination, that it would never rain again. Buzzards lay on thermals with wide brown wings scarcely moving all day, and magpies crossed the flock with that queerly uncertain flight, resting cheekily sometimes on the broad back of one of the Suffolks, to delouse it of ticks perhaps. Sometimes lizards would zigzag off through low forests of bracken stalks to escape my wheels and whirring blade, so mightily frightening to them, so almightily puny to me. It was

June, yet for some reason mushrooms had sprouted every-where, at least two months before their time. All the time, as I mowed and Phyl scythed, we wore a dark halo of flies about our heads, slow-witted and persistent and aggravating, with occasionally a solo effort thrown in of a horsefly's sudden prick on the brown arm's flesh, and often, according to the humidity, incessant attacks by myriads of invisible gnats against which we had no defence except flight back to the house.

This was high summer. We celebrated the Jubilee in labour, clearing and smoothing the bottom half-acre of Lower Rough as an area to be sacrificed in next year's lambing and naming it Jubilee Paddock. Golden Boy rested proudly, one supposed, among his matrons, whilst lambs gambolled in gangs, many of them proof of his own potency, some of them proof of old Sam's potency, maintained even in his last few months. For the first time all the sheep were free of foot-rot, a further memorial to Sam, for thereafter we were seldom to know a case again, and we could not but look back and blame all those hours of work upon Sam's old black shoes.

Any minute now the first sufferers would go down with blowfly strike. Whatever satisfaction a walk among our summer sheep might bring, it was always made imperfect by the knowledge that we still had no means of dipping our own sheep, and in particular that we could not dip the lambs and thereby protect them for the hot months from the scourge of strike. Without the dip, all that we could do was to walk suspiciously into the flock, to stand and watch for a lamb which did not play, did not graze, did not suck, which had lost interest in life except for an occasional lifting of its head from a listless pillow to look backwards to its tail in a helpless indication of the intense irritation it must feel back there where a horde of writhing maggots fed and thrived hidden in the root hairs of the fleece. Treatment, at least, was most efficacious. First-catch-your-lamb, less easy each day now as lambs grew stronger and we grew older. It always ended however with the satisfactory sight of dead and dying maggots dropping from the wool as the tar-smelling maggot oils drenched the affected parts.

"Poor thing," my shepherdess would moan quietly to herself as she parted the wool to make sure before lamb-release that there were no more hidden attacks. Only then would our puckered faces deregister the distaste we always felt. "Poor things. We really must get a dip, mustn't we?"

"Yes, of course."

Sheep had, beyond doubt, a penchant for disgusting illnesses. Orf could become even more unpleasant than fly-strike, making of a lovely lamb's face a hideous, leprous-looking thing and then spreading to the udder of its mother. Contagious ophtha!mia too was a ghastly thing if neglected at all, and I did not doubt that there were a score of others lying in wait for us, should we long neglect a daily keen inspection of the flock. All these troubles demanded individual treatment. Each case had to be seen, identified, caught, held and treated. A mere handful of cases of ophthalmia could take us the whole morning to treat, for treatment entailed squeezing ointment into the eye, and sheep do not really enjoy even being held, let alone having their head held still whilst somebody messes their eyes about.

This was supposed to be the easy time. Yet day after day they fetched me from my garden. Shearing, worming, moving, treating, marketing, day after summer day they possessed us, dragging us across the grass, messing our clothes, filling our minds, consuming our hours, never leaving us alone, wearing our lives down to the bone.

For sheep always had an urgency to them. A garden was gradual. Only the spectacular growth of June grass shouted its urgent demand that it must be cut now, but all else would sigh and wait quietly enough until there was an hour or two for it. Rosebushes waited demurely enough half a season for pruning and even days for deadheading or spraying; weeds would wait a week with never a nudge. But sheep, they were always now. Sheep-day by sheep-day, the sheep calendar rolled over inexorably from shearing to marketing to flock-making to dipping to tupping to wintering to injecting to lambing to shearing again, and all the time they wanted you there shepherding right along with them. I was tired every hour of every day and even more tired next morning. When occasionally I paused to peer into the mirror I saw an older face watching me. As for the beautiful and fastidious daughter of the one and only Charlie Kirby, her lovely hair was whitening, a tired paleness surrounded her eyes, and she trudged through her days with exactly that same suggestion of a brave limp to her right wellington as he had had.

Her pockets now were filled with fragments of hay and the detritus of sheep. Coats and trousers became stained with antibiotic-purple and her kitchen became lumbered with syringe-needles and feeding bottles. She felt their labour-pains within her own womb and she bore the sadness of a dead lamb more heavily than any word could tell. We fell asleep in each other's arms and we woke at morning again with thoughts of sheep seething through our poor addled brains.

The sheep of Dolgwili had us both in thrall.

24 Once the ponies have gone

All through April we had bandied about the phrase 'once the ponies have gone' as if their going must herald the onset of Dolgwili's Golden Age. It was one more retreat, as regrettable but as inevitable as my retreat from the vegetable plot had been, but the worry, alarm and threat we felt as they charged and turned and whipped about was no longer to be endured. They had always run wild on Dolgwili, and very wild too when springtime urges or boredom sent them into sudden galloping charges across the hillside, Fleur leading the race with wide nostrils, pointing hooves and flaring mane and tail, copied by Mushroom, Syndod and Mahogany as they all scattered sheep right, left and centre in that explosion of high spirits. Far below the Steep I would watch them, trembling at their mad gallop across that fearful obstacle course of rabbit holes, gorse bushes, ant hills and steepness. The donkey would caricature this charge in a braying canter along the lower and more gentle slopes at half speed, thereby confirming our belief in her superior intelligence.

The truth was that we simply were not true pony people. We seldom stabled them, never trained them beyond the halter stage, and although we had always hoped to have them broken for riding by our children and grandchildren, Simon, Trudi, Hannah and Ben each in turn came to verify by their lack of interest that truly we were not a pony family. If Charlie Kirby was looking down on Dolgwili, he must have been sorely disappointed that heredity had failed in this respect, for he had been a great pony man in his day.

Just as we had needed a door-opener to usher us into the sheep world, so we had needed one to open doors for us into the Welsh pony world. Owen Lloyd was a mature man of oblique comment, and when, having opened it for us some years before,

195

he was asked by us now to re-open the door to allow us out, he readily agreed. He even professed some obligation to us, as if he had known all along that we were not suitable, and further offered to transport our ponies to the first official sale of the season.

"Don't forget, give 'em plenty of oats now. Feed 'em up," obliquely he advised us, and for all that month before the sale, we had fed Fleur and her daughter Mushroom as well as we knew how. Mushroom, the foal who had fallen out of our meadow when only hours old, had grown disappointingly coarse-headed despite her good breeding. Fleur, aristocratic and dominant, had always been the one Phyl had pointed the finger at. These two certainly must go. The quieter Syndod and her yearling filly Mahogany we would deal with later.

Even at first sight the market appeared touched with bleakness. As we walked from the carpark through the little town to the sale-ring throng, faces seemed peaked with consternation. At the sale-ring the hum of voices behind the auctioneer's strident solo was subdued yet somehow rife with grumble.

The weather was fair but windily cold. We had travelled a long way, and during our sub-committee meeting in the warm car Phyl had surprised me with the backpedalling proposal that we should not let Fleur go for less than the two hundred pounds we had paid for her three or four years ago.

"Really? Surely that's optimistic? I should have thought we must let her go for whatever we can get and I can't imagine that would be two hundred."

She had pouted and sulked a little in her enigmatic way – as I imagined – but when I remembered her wintertime fears among ponies, and the day grandchild Hannah was found wandering among mercifully quiet pony legs, I knew quite well our pony venture was definitely over.

We listened to the bidding awhile. It had no life. The auctioneer would urge, chide, inform and beg, but his words dropped dead. As each cob, hunter, nag or pony was paraded round the sawdust ring and each detail of the proceeding impressed our eye, ear and brain, we became aware that the performance was almost a duet between the auctioneer and one buyer.

Then, "The butcher again . . ." we overheard a woman comment resentfully to her neighbour, and soon the next horse and the next would be knocked down to the same man. All was explained.

Clutching our catalogue we went off in search of Fleur and Mushroom. They stood haltered and nondescript in pens next door to each other, not under cover. For all our 'plenty of oats' they now looked poor among horses which must have been stabled and fed well all winter. On their flanks were stuck the white discs, red numbered, which marked them as no longer dignified by our ownership and almost doomed. They too seemed low in spirits; a useless anger began to seethe in me.

We carted about the market until we found Owen Lloyd in Welsh conversation with a ring of other breeders. A great cobby horse, jet black and bonny, was being knocked down to the butcher for £310 and the crowd-murmuring went on as it was led out and the next bright little pony was led in by an already anxious teenage girl in jeans.

Phyl's eyes were dull with despondency and my own face, I suppose, flickered with outrage as we listened to Mr. Lloyd's reading of the marketplace, of the new economics of pony breeding and the collapse of the Welsh pony market. Beautiful pedigree stock bearing proud prefixes of breeders renowned throughout Britain was going for slaughter or being withdrawn only to prevent slaughter.

The teenage pony-girl was engaged in hurried, worried conference with her father, whilst the grey-haired auctioneer peered over his spectacles from his rostrum, awaiting their decision in unfamiliar silence. Finally the unhappy girl mutely shook her head and, eyes downcast, led pony and father out.

Soon Mr. Lloyd took my arm and drew me aside. However obliquely he spoke now, he and I could avoid the life-or-death sentences no longer. Soon he would lead Fleur into that ring.

"But she's within a month of foaling," I protested.

His eyes dropped. "Yes . . . Yes, I know." He was a compassionate man, of great experience.

"By your best stallion," I added.

He nodded. We stared at each other.

He put his cigarette-laden fingers on my sleeve and said a little painedly, "Look, Mr. . . . " He never could say my name correctly and to save him I interrupted.

"She is not to go for meat," I stated.

"Well, then, what are we to do? What do you want me to do? She hasn't got much fat on her. That chap may not want her. I don't know . . . Have you got a reserve in mind?"

Fleur was to me classic pony beauty. She had a lovely head with bright lively eyes, she was as shapely as Grundy and had long bestowed much-needed elegance on our rough pastures. But . . .

"Look, I must sell her, but I won't sell her for meat. Sell her cheaply or get as much as you can, so long as he does not go for slaughter."

"Right." He hesitated. "Look here, I tell you what. If it comes to it I'll buy her myself. I don't want her, I've got too many now, but rather than let him have her, I'll do that."

And he did. From where we were on the outside of the ring we could not tell who was bidding, but first Fleur was sold for seventy and next her daughter went for forty-eight pounds. Owen Lloyd had himself bought Fleur, and an acquaintance had bought Mushroom, both to be kept for breeding. We paused only long enough to pass our thanks and then we fled that place.

Until then I had not fully apprehended how vast the difference was between a good sale and a poor sale. Everything drooped that day. We had seen anxiety on every face and tears on more than one, as family favourites bowed the knee to inexorable market forces, to the fundamental law of supply and demand. We drove quickly home in shocked yet relieved silence. How different it had all been at the start.

I remembered that hectic first day when the bubbling Owen Lloyd had rushed us from place to place in his expensive car to view ponies, more ponies and ever more ponies. He had been given his own first pony by a grandfather at the age of five, had been accumulating and improving them ever since, keeping them in little fields he somehow acquired in parallel. Forty or fifty ponies we had viewed that day, and finally, hungry, thirsty and with ponies-before-the-eyes, been so bewildered by Owen Lloyd's pedigree- and prefix-quoting enthusiasm that we bought the chestnut Fleur for two hundred pounds and the much poorer-shaped Syndod, a younger mountain pony, for eighty.

It seemed to me that something in Welsh blood responded to the pony just as it responded to song. You could almost see a different excitement nerve the Welsh at the sight of a fiery mountain pony or sweet-moving cob, a broader smile, a brighter eye, a shout and a leap of *hwyl*. A good-moving pony could excite anyone, of course, but not with the same fire as in Owen Lloyd or, soon, Cardi Williams.

We had made our original entrance to that pony world most correctly. Owen Lloyd had been recommended by the Society itself; he was a very well-known breeder, an official and a show judge. We became paid-up members, lapped up all the official advice and annually attended the yearly stallion show.

On our very first visit to this area, whilst still in search of a new home, we had fallen among pony folk at a chance agricultural show in a small town. Standing there amid such friendliness, I had gazed out from the busy little show to green and sheep-laden hills all around, tree'd in patches and quite unspoilt. I thought how desirable it was to live near such rural calm with such friendly folk and that first impression long remained.

The stallion show at which we were to choose a mate for Syndod was a first cousin to that original show and held on the very same field. From all over Wales exhibitors had assembled the very best stallions and their progeny in competitive classes to guide such as ourselves towards excellence.

The long hot summer had already started, even in April. The field jostled with countrymen in best clothes, with stalls, horseboxes, Land-Rovers, cheerfulness and red faces. It was a day out in worship of the horse. Ponies and cobs stood or were being led about everywhere, glistening, hard with springtime fitness, plaited and oiled, dolled up to the nines and beyond.

Owen Lloyd was exhibiting Merlin, the grey stallion we had already used on Fleur, and whilst he groomed Merlin's perfection to further perfection beside his trailer, we sounded out his opinion on a stallion for Syndod. ('Sundod' it is pronounced, meaning 'surprise'; she had surprised Owen by the earliness of her birth.)

Even to our inexperienced eye, Syndod was less than perfect. Her back was poorly shaped and the set of her tail was noticeably low. We must seek a stallion to correct these faults in her offspring and soon Owen Lloyd pointed us towards a young prize-winning stallion called Twyford Dolphin, owned by a well-known breeder called Cardi Williams. We left our mentor silkily caressing the magnificent Merlin, tossed him our thanks and wishes for success in his class and, pony people at last, sought Cardi Williams.

Judging had started in all three roped rings. Tannoy announcements rattled across the crowds, young mothers in inappropriate shoes pushchaired ice-creamy babies across the grass, expectant-faced young men walked their ponies, horse-boxes crept through the throng. Cardi Williams, our programme told us, was due to show, not Dolphin but an older and well-established stallion called Assington Gremlin, winner of almost everything over the past four years.

Fifteen pony stallions were already lined up in Ring Three as we joined the spectators. Stewards marshalled, owners fussed, the bowler-hatted judge hovered, spectators stared. All waited. Over the loudspeakers again went the call for Cardi Williams. Faces turned in impatient search, the judge consulted his stewards.

"How on earth can anyone choose between these," Phyl commented wistfully, gazing at the magnificently turned-out creatures. Their manes and tails were coiffured to advertisement perfection, their heads were proud and high, their muscular bodies shone with fitness. They were the loveliest animals in the world.

"We'll pick out one each," I suggested. I knew I should pick solely by colour.

But at our backs, rumpus was developing, shouts and the banging of hooves against wood, in a large horse-box parked about fifty yards away. Suddenly the milling crowds began to fall back hurriedly as a chap dressed all in white, leading a pony dressed all in white, charged down the ramp of the horse-box with warning shouts and cut a relentless swathe through the crowds like a white harvester through a standing crop. Onwards it charged amid more shouts and laughter towards our ring and into the ring and all round the ring, electric and magnificent, the powerful little white stallion at peak life and full canter, mane and tail flowing, hooves thumping the turf, knees high, spirits higher, and, alongside, young Cardi Williams just as high himself with sheer dramatic impact, running in step with his beast, his right arm out stiff to hold the stallion's head, his knees exaggeratedly raised to match or to instruct the pony's knees, his face red-hot and curiously still, pointed like a solo dancer all intent on exact physical performance.

Welsh pony-excitement breezed through the smiling crowd, fanning their appreciation into shouts of encouragement, until Cardi began to slow and pull Gremlin to a mere snorting walk and, a good boy now, assumed his place in line with the other waiting ponies and allowed the judge to proceed with his formal and deliberate examinations of conformation and movement of each pony towards that final and – it seemed to us – inevitable moment when Assington Gremlin and the sweating Cardi stood top of the chosen few to receive his red rosette.

As Cardi and Gremlin set off again on a splendid lap of honour, bursting into a trot with that immediate full power which suggested that Gremlin might take off with the struggling Cardi, our applause did not perhaps have the huge spontaneity the winner deserved. Instead we clapped moderately and smiled almost ruefully, for the judging had seemed almost a charade; had we not known all along that there could be only one winner after such an entrance?

"Still," we told each other, "we didn't do badly. Your chestnut came third. What about Gremlin? You fancy him?"

"I could do."

"White, of course. What would white on dun produce, I wonder."

"Let's find Cardi and talk to him."

Talk to Cardi as he rested on his laurels by his horse-box was instructive fun, straight from the horse's mouth. He was handsome in a ponyish sort of way, cheerful and easy, and he thought the best thing would be for us to go over to his farm, see his ponies and pick our stallion.

"I've got nine to pick from," he said modestly.

"Nine?" we carolled. "Nine stallions? How many ponies have you got then?"

"I don't know for sure. About a hundred and thirty," he joked.

To us, Cardi Williams was superman. Where I was middle-aged, timid and English, he was young, bold and so exuberantly

Welsh. The joke about him having a hundred and thirty ponies was no joke. He farmed six different farms, he told us when we accepted his invitation a week or two later, had one hundred and twenty-eight ponies at the last count some time ago, ran a flock of six hundred sheep, a barley beef unit, exhibited at all the main shows throughout Wales and England, was courting a horsewoman who lived in Surrey and accomplished all this with only the help of a tractor-driver, part-time, and a sister, part-time. I would not have been really surprised had I later discovered that he regularly skied in the Cairngorms, produced the village pantomime, sang with Treorchy Male Voice Choir and was a Welsh 'B' trialist at wing-threequarter.

His home farm stood high, about twenty miles from us, within a sniff of the seawind. Small pastures sheltered among forestry and the whole farm gave the impression of being unchanged – apart from new buildings – and unchangeable, a family farm with a stud of ponies founded generations ago.

The stallions were stabled near the farmyard. Gremlin looked comparatively homely now, off-stage and without make-up, and was housed, not in the posh rows of stabling which we might have expected for so famous a stud, but in old barns and stone stables, all quite without pretension.

Neither Assington Gremlin nor Twyford Dolphin were really available now, for they would be travelling to shows, but he did have a very nice new stallion called Mahog, of the same Twyford strain, dark chestnut and unproven as yet.

Cardi led him into the yard and, with a slap on his flank, sparked him into clatter-hooved movement. We explained about Syndod's low-set tail. Mahog, Cardi assured, had a very well-set-on tail. Okay, we said soon, very pleased, Mahog it should be.

Cardi himself seemed keen to show us all one hundred and twenty or thirty ponies, and despite my muttering that surely he could not spare the time, he threw us into the front seat of his Japanese estate car among the halters, branding irons, old ball valves and the chainsaw, and set off like a rally car out of the farmyard, through gateways, up rutted lanes and out on to hilly pastures where grazed his cattle, ponies and sheep. Occasionally we stopped violently to get out and inspect a cob mare soon to foal or in season, but for the most part we rally-crossed bumpily across country, Phyl and I metaphorically clinging to each other in terror, whilst Cardi gave us a commentary, a guided tour through Cardiland, and ponies stared at our fast-car invasion. Evidently this was a regular progress of the king through his wild kingdom, for gates stood fixed open ready for him. Sometimes he had either forced or worn an opening through banked hedgerows and I half-expected to encounter a few water-jumps. Only twice did I have to get out to open gates for him, and that only when we came to sheep. He had just finished shearing his six hundred sheep single-handed.

"Chap wanted seventy quid to shear them and I wasn't paying out seventy quid when I could do them myself," Cardi explained whilst I shuddered to think of all that labour, and recalled Haydn's joke about counting your fingers after you had shaken the hand of a Cardigan man.

Never mind; Cardi had bounced into our true affections with all that he so energetically did, and we arrived back in the farmyard dazed but easy winners.

"Now, transport. We have no pony transport. Shall I hire somebody to bring Syndod over?"

Cardi thought.

"When will she be 'in' again? Have you noticed?"

"She may be 'in' now. The donkey is mounting her, or was yesterday anyway."

"Are you sure? The donkey was mounting *her*?"

"Yes."

"Well then, she's in. I'll bring my stallion over this evening. Okay?"

"Fine."

We drove home, settling after the hurricane. Sure enough we saw his horse-box pull off the road at mid-evening. Within minutes Cardi was leading the dark young stallion down the ramp and through on to the bridge. For some impetuous reason he there decided to ride her, bareback of course. He tied her to the bridge rail, stood on the stone rampart high above the river and tried to mount. The unbroken pony plunged and reared whilst Cardi slipped and hung through various attitudes of discomfort, before trying again and obstinately again before final quittance as I arrived.

We consulted.

"Can you bring your mare down here on to the track? Will you be able to hold her?" he asked doubtfully.

Seldom can the aged resist the challenge of the young. Syndod led well. She came sedately down towards the farm gate, the other side of which Mahog plunged and snorted in a playful and dishevelled fury of action, which must have delighted Cardi as he hung on to the rope. With some preliminary skirmishes the four of us accomplished a rough and ready introduction there by the farm gate. Initial desire, however, soon waned between Syndod and Mahog and, after a little further pushing and pulling, Cardi decided that probably Syndod had gone past her time.

To save another journey he decided to take Syndod back with him. The chance that Mahog might mount her en route could be nullified by judicious swerving and braking tactics, he thought, and so we loaded them up. Hooves thudded and ponies whinnied as we closed the ramp on them, Cardi jumped into the cab and drove straight out into the holiday traffic and I watched them disappear, aware of a strange peace settling about me now.

I phoned him every now and again.

Oh yes, she had been mated all right. Keep her there just to make sure, eh?

Oh yes, she had held. Sure. Yes, he'd be coming this way next week, he'd drop her off.

No, no, she was all right. She was on a lovely bit of pasture down by the river, looking a picture, she was. Lots of herbs in the grass down there. She was twice the pony now.

The next thing was that he could not catch her. She was running with four other wild ones and had gone a little wild herself. He would bring her back just as soon as he could get hold of her.

Dolgwili's long hot summer passed Syndod-less. It did not matter.

Then, one night in September, when we had an Australian family staying with us, the Aussie father called down to us from the bathroom that some headlights had just pulled in off the road and were shining up towards the house. It was eleven-forty-five. (The Aussies were obsessive bathers; the four of them took three and four baths each per day, shamelessly even boasted of all this cleanliness, whilst I dumbly curled up in my own filth and trembled to think of our poor drowning septic tank out there.)

I went out into the sultry night. It was Cardi, leading the long-holidaying, blossoming Syndod.

Well, in due course Syndod begat Mahogany, a real little beauty of a filly who represented the high summer of our pony adventure. From then on the days dwindled down to the wintriness of that awful sale.

"That, anyway, is our first and last pony sale, my love. Never again. Either we will keep Syndod and Mahogany for ever, or we'll give them to somebody who can guarantee to keep them."

I thought about Cardi and his ponies. Market collapse would not worry him unduly. Ponies and the Cardi Williams of this land were not about money, about prices; they were about life. So long as there were Welsh hills there would be Welsh ponies, and wild young Welshmen.

25 He was his grandfather's grandson more than his father's son

It was nine o'clock in the evening when he came. He stood inside the kitchen door in turned-down wellingtons and round his neck he sported a red-and-white neckerchief. He smelled of open-air and the permanent underlay of cow-smell which he must have inherited from old Charlie Kirby. Although he was a son in the home of his parents, something in his manner suggested that it was strange for him to be standing here like this.

The saluki and the boxer, body to body, stirred in their stoveside comfort.

"That coat," Simon quietly commented, considering Caramelle.

"It's the operation. Like me, she's lost her looks."

"Old Sam isn't well either," Phyl said. "He yelps with pain in frosty weather. I suppose we'll have to have him put down some time this winter."

The three of us observed the old dogs with scant interest. Show dogs, he called them spurningly. Working dogs were his love. As he stood there, tall and too thin, I realised he was as lurcher-like as once I had thought myself saluki-like.

Caramelle sprang up, suddenly alert, a growl rattling suppressedly in her teeth.

"Have you got your dogs out there?"

He nodded. "Coming out? I've brought the lamp."

I sighed and reached over thoughtfully to stroke Caramelle's long and silky brown ears.

"I still regret we didn't mate her with your Urchin," I said. Every time she had been in season she had lived and slept alongside Sam-Boxer, but she would never allow him to proceed with his somewhat nominal wooing of her. Urchin had been different. She always fancied Urchin, however 'kelb'.

"I've already been out with the gun, earlier, at dusk," I answered at last, hinting at my tiredness.

"Get any?"

"Didn't have a shot."

Simon stirred slightly, intimating impatience to be out there in the night. "Coming?"

I smiled and shrugged. "Yes. All right. For a little while."

I caught Phyl's eye. It was slightly amused, I think. "Wrap up," she commanded me.

Could he not see how I had aged, how tired I had become, how pleasant my chair and the warmth here? Yet he was this strange thing, my son. And he wanted me to go with him into his world, wanted me to see his delight and the power he held out there in his kingdom. So I fumbled my way back into my boots, scarf and coat and followed him into the cold night air. Three dogs instantly came padding up to him, two lurchers and his springer spaniel. He pulled out several pieces of blue baler-twine from his hunting-bag, still in loop form from the hay bale, and slipped them through the collars of the lurchers. We set off in untidy convoy through the dormant garden.

I wondered, as we drifted, whether he had leashed the dogs in deference to the unease I always felt when dogs were loose on the farm. Even now my mind harboured reservations on the general wisdom of dogs hunting on Dolgwili, and as though he sensed it, he stopped at the gate into Sheepdip and whispered, "No loose sheep?"

"No. All across th'other side in Jubilee Paddock."

We went through the gate, silent with the latch. Automatically I check-listed the animals in my mind. Syndod and Mahogany, the donkey, the hens, the goats, all the sheep. He made small whistling commands to keep the spaniel close as we mooched across the pasture and I had to hurry more than I wanted to.

It was a clean cold night. Drying winds from the east had scoured the grass for days and the great night sky looked polished by them too, and cleared of cloud.

We came across Sheepdip to some electric netting set up to save grass for lambs not yet born.

"This on?"

"No."

We stopped there on flat ground and he hoisted the lamp free from the bag, slipped the leash from the lurchers and switched the lamp on. The power of the beam surprised me as it sliced through the darkness and across the ground for a hundred yards. Methodically he directed it to search the whole meadow and rabbits began to run. Three or four . . . six.

"Go on," he urged the lurchers, choosing one rabbit running in the beam.

The lurchers were gone invisibly but the rabbit had too much of a start. As it disappeared into furze, another got up and fled, and dogs and beam switched to that, chasing along the contour of the slope. For a moment that one too seemed to have escaped into brambles up near the track, but then the lurchers had it out from there and were closing on it when the rabbit ran straight into the wire-netting fence of the new orchard. The dogs had it instantly.

"Go on, Zip, get it," Simon commanded the spaniel at his heel, and broke into a trot himself.

I strolled after them in darkness. I could hear the rabbit squealing in distress. He would be there by now, snatching it from the squabbling lurchers or receiving it from the soft mouth of the spaniel, lifting it to despatch it with the sharp edge of his hand, dropping it into the large bag.

"Try up the Steep," he commanded when I arrived.

I followed him through the hedge on to the track, trying to walk silently, trying to dissolve into the night for his sake. He was a different man out here; confident, expert. Out here we had changed roles.

By the Spout – there was hardly a tinkle of water this evening – we crept to the earth-slide formed by sheep to get through the hedge. There he switched the lamp on again, beaming it upwards to sweep the Steep. Three rabbits bolted in different directions and the lurchers were off again.

The lightbeam ran tangentially off the earth into the vagueness of air and sky as the hill sloped, and perhaps the lurchers were chasing towards the stars, for they were out of sight when we hauled ourselves up through the hedge gap, him easily swinging, me heavily heaving. He went ahead at his trot and I did not doubt he was the hunter tingling with the excitement of the chase, quite oblivious of the physical effort of running uphill over uneven ground, whereas I was no part of it all.

The incline here was carved narrowly and unevenly in footpath lines by sheep in regular daily progress, or was swollen suddenly bulbous by the enigmatic lifestyle of ants. Infant brambles had laid tentacles across the mossy turf and twice they almost tripped me as I walked the ever-changing earthangles towards the centre of the Steep, where existed the year-after-year rabbit burrows now occupying the attentions of man and dogs.

"Well? Did they get another?" I dutifully asked, arriving short of breath.

"Yes."

There seemed to be a lull in the hunting. He switched on the lamp again and swept the beam across the hillside but nothing moved now. Everywhere the dark brown of rotting bracken smudged the ground, and the wind blew coldly into me.

"Did I tell you I came across a gang working this hole about a week ago?"

"Who?"

"Don't know."

"Three blokes with ferrets?"

"Yes. And terriers. Guns."

"A Sunday?"

"Er . . . Yes."

"Boys from the valleys. Shot a sheep on Panteg."

"What on earth for?"

He allowed the unanswerable question to blow away into the night. Absently he played the beam in all directions.

"Say anything to them?"

"No. I was down there on Sheepdip at the bracken. I couldn't bother to come right up here. Besides, I want rid of the rabbits."

"Just as well. Tough, those valley boys. Miners." He switched off the lamp. "Try over Thomas'," he decided and began to climb immediately, straight up. I hesitated before I began to follow him without enthusiasm, testing my thighs against the steepness. They did not really want to go, but I forced them to take me along a slow zigzag up to the shelf at the top where he waited.

When I had recovered wind, and feeling that he was about to

assault the root-tangled bank and the hawthorn hedge atop of that, I said, "I think I've had a fair day. Do you mind if I leave you to it?"

"Yes, all right?"

"Just tired. Another night."

"I'll call in."

Relieved of me, he moved with extra impetus it seemed, hauling himself upwards on an old sawn ash-stump and somehow over the hedge with the dogs nosing a way through the hawthorn. They disappeared. I began to descend. Recall of a film was in my mind, from Bronowski's *Ascent of Man*. A nomadic tribe, herding their sheep and goats along traditional routes which they had followed for generations, had a certain difficult river to negotiate. Those old men of the tribe who this season could no longer summon strength or will to cross the river were left on the riverbank to die.

I sat down.

Below me stretched my unseen kingdom. All that I owned, all that I had achieved, all that I was, lay spread on the twenty-acre invisible carpet which fell sharply away from where my boots rested. Down there in the warm cottage – even its lights could not be seen from up here – she would be preparing the supper tray, stoking up the fire, listening to the dreadful news. The old dogs would be sleeping by the fire.

Orion looked magnificent. I remembered other nights, other skies. I had known all these once, Castor and Pollux, Betelgeux, Sirius, all that lot. Yet now, when I regarded the mystery, I could name only Polaris, Orion, a few like that. It did not matter any more. Courting nights I remembered, under these stars, the long arm-entwined walks along country lanes watching for falling stars, arguing about nothing, and all that kissing. Lovely innocent nights . . . And this, our son out hunting under the same stars, this was the other end of that kissing. Had I told her then, or had she told me, that we would finish up keeping sheep on a Welsh hillside, how impossible that would have seemed.

Well, if the purpose of life was life, we had fulfilled life's purpose. I had played my games, fought my war, won my woman, established my territory and a place high enough in the pecking order to satisfy my modest ambitions, and – most important of all – passed on my genes. That I had preferred to do all that in the company of animals and in the countryside rather than in the company of humans in cities was merely a matter of temperament.

This here, this now, this was all there was. These unknowable stars, this cold air chilling my nose, the sleeping flock down there, the warm home waiting with wife and marriage bed, the family, what more was there to desire? Who on earth wanted immortality, who in heaven required life everlasting?

I began to descend. This was by far the best winter we had ever known, up till now. The worst onslaught was yet to come, no doubt, but at least we were already fortified with a shed full of hay, for so far were we taken over by sheep that what had always been thought of as a garage, was now thought of as a hay shed. Every other day the little car fetched up four bales, two to go directly into our new hay rack over in Jubilee Paddock where the flock was permanently installed from now until days before lambing. It was a joy to serve them, for no sooner was the rack filled anew than the ewes gathered round in orderly fashion, eight each side, nosing and daintily chewing the sweet hay without waste or worry. When they had consumed their pound or two they would move off to sip from the stream, or to comfort themselves down on to the earth to chew the cud. They did not attempt to break out, nor did the donkey and two ponies attempt to break in. The ponies had not been put to the stallion, they had hay and they had shelter. All the animals had never looked so well. They throve on hay-contentment and it spread to us.

No, I told myself taking my time down the hill, it was not that I no longer felt the promptings which instinctively set men racing for glittering prizes or some sort of immortality; it was simply that I did not heed the promptings any more. I was nobody, nowhere, a mere minder of sheep. And that was enough.

I was cold when I arrived home.

I was pouring my last cup of sweet steaming tea by the fireside, Phyl was sipping her last coffee, when he called back in.

"How many then?"

"Eleven."

He stood in the doorway between kitchen and sitting-room,

not quite at ease still. He never took his cap or his wellingtons off. You had the impression that the wind was still blowing about him or that he had brought the outdoors in with him.

"Do you want some tea?"

He hesitated and I sensed he was preparing to say something which required daring from him.

"Please. I ought to be getting back."

He nearly said, "I've got a wife and kids somewhere."

"Good, wasn't it?" I said, pouring tea. "Eleven in an hour or so?"

I handed him tea. The porcelain cup looked totally inadequate in his mittened hands.

"Caleb and I are thinking of giving up our jobs to go rabbiting for a living."

"Good," we both exclaimed looking quickly in surprise at him from our chairs. He smiled that beautiful smile, relieved perhaps.

"Get fifty pence a rabbit. Got a bloke from Birmingham will take all we can get."

We discussed it a bit, the risks, what did his wife think, what about myxomatosis. He worked long agricultural hours at awful jobs for poor pay in atrocious conditions. By vocation he was a hunter, or gamekeeper or poacher. He had been able to snare rats with piano-wire when he was twelve. He inhabited a hedgerow world, knew every footprint and every bird, could track and kill, could train dogs, so what on earth was he doing spreading slurry, milking Friesians, being manservant to porkers and battery hens?

More than his father's son, he was his grandfather's grandson. The genes had skipped a generation. Even to the dogs. Charlie Kirby had been a great man for his greyhounds and coursing. Perhaps it was some unconscious reverence for his grandfather that was the fount of his continual admiration for old countrymen.

We gave his new rabbit-career our blessing.

"Be up Sunday," he said, departing.

"Good. Goodnight."

26 Most admirable animals

Of the original Beulah Speckleface which poor Thomas Dewdrop had contrived eventually to buy, at last gasp almost, for us
at the Llynwalter clear-out sale, eight ewes still flourished
sufficiently four years later to be in our sacrifice-paddock to
prepare for another lambing. They must all now be eight years
old at the very least. Some of them would be what we had learnt
to call 'brokers', meaning broken-mouthed or toothless, and
others would be 'full-mouthed', meaning veteran but having
their teeth. If in our moves towards that healthy-looking and
highly performing even flock of young sheep which was supposed to be our ideal, if in those moves we sometimes felt that
the Ancient Eight ought somehow to be disposed of, our gratitude and loyalty to them quickly shamed the notion, and indeed
even the economic arguments told us we must keep them, for
replacements now were costing from thirty to forty pounds.
Further our favourite sheep-book recorded an ewe of twenty-
eight years which had bred twenty-four crops of lambs and was
still breeding, and on the question of how long an ewe should be
kept, recorded a wiseacre's answer: "Until two days after she is
dead."

The Ancient Eight were to us, élite survivors and favourites
and most admirable animals. Indeed the affection which both
of us felt for sheep in general had been mostly won by this eight.
They had spoiled and interfered with my most cherished
gardening ambitions, there were times galore when I cursed
their demands upon my time and strength, yet their bravery,
their hardiness and, I think above all else, their uncomplaining-
ness, the way they accepted pain, discomfort and all the rigours,
these were virtues for which they had won my undying regard.

If, as I had come to believe, we are all of us animals – and although so many folk use the epithet 'animal' as one of contempt, I have come to feel touchy when they do so – then it seemed to me that the modest and seemly way in which sheep conducted their lives was far and away a more admirable way than that in which the human animal conducted his. If I could claim on their behalf no sheep-Beethoven, I could point to no sheep-Hitlers, and to take the analogy no further, it seemed to me that they used the earth to live harmoniously and with dignity in a way that we did not. I had lived long enough, I felt, to indulge myself in an affectionate admiration for sheep but not for my fellow men. Sheep were better.

And no sheep was better than old Darkie. Right from the start she had been a distinctive sheep, and that for reasons other than the darkness of her fleece which had given her her name. She had always been a loner, and perhaps the greatest survivor of them all. Just a year ago she had virtually been in the condemned cell, as it were, and that not for the first time. Last year she had suffered prolapse of the cervix before lambing, had lambed without help and raised twins without trouble before amazing us by revealing, when the time came to make up the flock again in August, that she was fitter and in better all-round shape than she had ever been before. We scored her at two-and-a-half with no difficulty and, with her survival record in mind, kept her with the breeding ewes. In other years she had seemed rheumatic in her hindquarters and a little eccentric in her mothering of the twins which annually she produced. She seemed to forget them at times, and more than once we had come across small dark twins at dusk tucked cosily into the shelter of the fallen elm or of some tufty growth, displaying as much independence and unconcern as did their mother perhaps half-a-mile away. For Darkie herself was the scavenger par excellence. All sheep are reckoned scavengers, but none scavenged as thoroughly and conscientiously as Darkie. She prowled quietly and profitably around, finding food without fuss when other ewes were running crazy past the very food they were demanding. With her greatly improved condition in mind, we supposed now that in other years she had never recovered from one lambing before she was going to the ram again. She had been terribly thin in those years, and no doubt her milk

214

supply was poor. This would have accounted for the regular 'desertions', as discouraged lambs gave up sucking at an empty udder and as poor old mum struggled to scavenge enough nourishment to make good her overdrawn account of metabolism. One year the twins had grown hardly at all and Simon sold them off as store-lambs to a friend who spent a Welsh fortune to fatten them. All that, however, we were now willing to overlook, so changed was Darkie's pre-lambing condition, although also, I privately suspected, so newly attactive was her natty dark-grey to black fleece to my shepherdess-wife's spinning ambitions. Darkie thus occupied her place in the sacrifice-paddock with our certain respect, still maintaining her condition with that slow prowl, her black nose to the ground sniffing to find every last oat or nut fragment, however muddy, ignored or refused by the rest of the flock.

True, Darkie lived with a question-mark hanging over her, but then so did I and so did several others. Black Collar was certainly the most broken of all the 'brokers', quite unable to chew the cobs quickly enough to win herself a reasonable ration. Her question-mark was darker for this, and we told ourselves that surely this must be her last season. Yet she had always been such a good sheep and had reared her twins quite successfully last year. What does one do?

Of all the old ones, Spotty was our undoubted queen. She was the friendliest and the fittest of all the Speckles we had ever had, always the last to lamb each season, always rearing twins – triplets once – and since she still had 'all her own teeth', she always looked well.

Five others – Dainty, Nice Fleece, Daisy, Cripple and a scary younger one, which we could only name and rename Wild One – these were all survivors simply because they were good sheep. They had survived all our ignorance, our incompetence and our foolishness over the years and rewarded us more generously than we deserved. If sometimes they had led us a dance, even these 'good'uns', through the hedges and across the awful places of Morgan's Wood, had made long and anxious our nights at lambing, or fretful the long winter days, nevertheless they were good friends of ours. We saluted them daily as we counted them, fed them twice, inspected and watched them, the doyennes of our flock.

215

Accompanying the Ancient Eight in the sacrifice-paddock this season were eight of their daughters, selected and held back two years ago as replacement stock, for those Speckles who had died or been too wild for us to manage. We called these youngsters 'gimmers' because that seemed the term for them, although the nomenclature of sheep seemed as regional as dialect itself. When we tried out 'gimmers' on some local farmer he stared at us uneasily until we explained in our Suffolk accents, and he came back at us and said that in these parts a gimmer was a yearling, or, once it had been shorn in its second summer, a shearling.

"All right," we agreed. "Our gimmers are shearlings."

Habit and unthinking being rife with us, however, we continued to think of them as gimmers – the Red Gimmer, the Blue Gimmer, and so on from the coloured tags they now wore in their ears. Many were the offspring of old Sam-Ram, and the eartags identified them quite precisely to Phyl as "Old Dainty's pretty little ewe lamb, don't you remember?" or "Daisy's big lamb, haven't you noticed how close they always are still?" For she identified with the flock far more intimately than I did, or it may have been that I was a little dim, or my head too stuffed full of 'hedera canariensis' and that crew for it also to be able to cope with Little Gentle White Tag and so on. In fact they were always her sheep rather than my sheep, but since I was her man they were *our* sheep.

These gimmers, then, were Speckleface-Suffolk crosses. Their size, and the halfway markings of their face, showed them thus to be and they bore all our hopes this lambing. They were home-bred and therefore of extra interest to us, and first lambers also, something we had no experience of.

The attachment we felt for our original Speckleface and these eight home-bred gimmers had never really extended to our pure-bred and expensively bought-in Suffolks. Times occurred certainly when the docile nature of the Suffolks compared so favourably with the breaking-out nature of the marauding Speckleface that we willingly transferred favouritism to them for a day or two, but soon their hugeness of body and appetite, and what seemed to us a proneness to minor ailment, persuaded us to transfer back once more to the unbeatable Speckles.

No doubt temporary favouritism would have become permanent if only the Suffolk lambs had been good enough to make profitable our pure-bred project, but, alas, it was quite other-wise. Our first two pure Suffolk ram lambs, which we had honoured in optimism with names, Plato and Hercules, had become soiled and unhappy looking creatures that never prospered as they should. The first Suffolk ewe lambs too, which we also kept, similarly disappointed and puzzled us. We kept them running with the flock more in hope than pride, and Plato and Hercules did become ram enough to chase the girls and so keep Golden Boy jealously on his toes.

The Suffolks, now enclosed with Speckles and gimmers and having no choice but to hobnob with them in this single acre of daily more sacrificed grass, never lost their separate identity on Dolgwili, never mixed in to become a multi-racial society. All through those days when the flock had a choice of grazing and movement, the Suffolks moved together in their one small group slightly apart from the Welsh natives. No more than their owners – who could scarcely not compare their own retained separateness and Suffolkness amid Welshness – did they seem to consider themselves superior, but certainly each group al-ways seemed to be as aware of differences between them as we ourselves did of differences between Englishness and Welsh-ness. How exactly a sheep decides she is a Suffolk and will keep with Suffolks I could never work out, but I assumed it was a matter of smell; Suffolk-smell must be different to Speckleface-smell. And Williams-smell must be different again.

A one-day course in sheep-education – totally informal and private of course – had been our attendance at a sheep market in a nearby country town, a very small town at that, where, could we believe it, eleven thousand breeding ewes and lambs were on sale. Arm-in-arm and happily anonymous under a splendid umbrella on which the rain thundered all day long, we wandered around acres of thickly-penned sheep, stopping, as the mood took us, to exclaim how nice this pen looked or how messy that lot, unhurried and quite without anxiety or the Englishness that had so marked us out at our first sheep sale way back at Llynwalter. By now, sheep half-owned us. Sheep-ness showed in our faces, and in our dress, and in our conversa-tion. We would pass certainly as sheep-farmers, dressed in workaday macs and wellingtons, and possibly even as Welsh sheep-farmers in this most Welsh of places, unless at last we must speak to somebody – when almost certainly we should pass only as Australian. This accusation, of us being Australian, had been levelled at us a score of times in Wales, so that I had devised the standard explanation that Australians were all descended, with their accents, from convicts, all of whom came from Suffolk. In any case this explanation would hardly be needed today, for we had nobody to talk to, no questions to ask; we were not here to buy, only to observe.

The market was cosily situated down at river-level, in a cul-de-sac well off the main through-road. The comings and goings of cattle-transporters, cars, vans, Land-Rovers and trailers were slow and continuous, a marvel of chaos and cheerful patience in which vehicles manoeuvred and re-manoeuvred, poking noses in for likely exits or entrances, trying round the other way or sitting waiting in cheerful silence for the man in front to move, to unload or to come back from the pub or sale ring, whilst men shouted good-naturedly amid this regular market snarl-up. Everywhere countryfolk passed on their business, their faces tilted down against the heavy rain, a milling, chattering crowd of buyers and sellers and watchers.

This market sprawl divided itself roughly in two on either side of an approach road, the cul-de-sac, narrowed by stalls selling wellingtons and spanners, fish and tea and horse-riding gear. Ewe lambs were being sold on one side of the market, simultaneously with ewes on the other. The whole thing managed to give the impression that it had no brain, no organising force, no regulating power. It seemed that it still bore the imprint of its origins generations ago when the first neighbours met here outside the pub to sell or hire, that from there it had just grown, with somebody deciding to put up a stall here, or an office there, or to park his cart over there. Traffic had no flow. They edged in, the motors, somehow, and they eventually edged out, somehow. The policeman was away fishing no doubt, and what few regulations had been made were now ignored. 'No Parking', a few poor old notices muttered unheeded and down here you drove on the left, the right or the centre of the road on which yellow lines had not yet arrived. The market, in fact, seemed an organic thing, something with its own pulsating life, whose cells were sheep, whose lifeblood was money. It was so lively, so friendly and cheerful, so Welsh, and as we drifted through the crowds with that happy feeling of being liberated which Wales had so often bestowed on us, as we listened to the prices or watched a farmer unloading a late batch or another stepping into a pen to examine teeth – always teeth, nothing else interested them – gradually we began to feel our temperatures rise with sale fever.

We did have, did we not, a little capital still left. We could keep a few more. Half a dozen, say? After all, our pasture had been limed and phosphated.

"What are they? Yearlings?" we found ourselves asking a man in a pen which contained six very nice Speckleface.

"Aye."

"Have they had lambs at all?"

"No. Aye, well, I didn't have a tup, see. I just kept them in the orchard, like."

"You're not a farmer?"

"No, no. Like I say, I just got the old orchard and I like sheep."

"They look nice."

"Aye. Oh, I've looked after them."

"You didn't rear them or . . . I mean . . . " I meant, where did they come from.

"No, no. I bought them as lambs last year. This sale twelve months ago, like. Aye. They're good sheep."

"Yes. Well, good luck. We may have a bid."

We stood back, pursed our lips, observed the sheep. This part of the mart was under cover. We were both thinking, "Six would just about go in the car. There aren't many lots of only six. How much?"

Doing private sums, we moved on. Buy them at what, seventeen pounds as lambs, put them in an old orchard somewhere and sell them a year later, at what, thirty pounds? Soon we found each chatting the other up on the idea of buying just six lambs, an idea which began to seem positively brilliant when a quick dash back found the auctioneer easily getting thirty-seven pounds a head for our part-timer's six pretty ones.

We were bad for each other at sales of all kinds. We rubbed against each other too easily, caught light and burnt ourselves through too little care. Very soon we were tramping around sorting out which pens we could possibly buy. The car would hold seven, if Phyl had one sheep on her lap. Out of eleven thousand sheep on sale, only five pens could we find which held seven or less, and of these five, three pens were occupied by lots respectively of eight, seven and five. These were all of a kind, small Speckleface-type from, or so we guessed, mountain stock. Very pretty.

"Oh yes," she said, pressing my arm. O Eve . . .

We lost the first two pens when the price soared over twenty pounds but won the last pen at nineteen pounds a head. We even hugged each other at this hugely transparent success, went and queued for half-an-hour at the mart office to pay for them, and then queued in the car for another half-an-hour to get near enough to load them up.

A limping cheerful man named Williams waited by the lambs. He put into my hands a one-pound note, the traditional luck-money, and a most gracious young Welshman stepped forward and lifted the five small lambs into the Renault for me. I had the impression that he had had his amused eye on us for some little time, that our helplessness or amateurism, or even the bravery of the little red car taking on all those Land-Rovers and transporters in the fight to get in, something anyway moved him to offer his help. It was most welcome, for even with small sheep it requires a certain athleticism to negotiate iron hurdles to carry a struggling animal into a car not actually designed as a sheep-transporter.

"You're a long way from home, aren't you?" he smilingly asked as we locked the tailgate on our precious load.

"Ooh . . . er, yes." I wondered whether to fetch out my Suffolk-convict-Australian story or to try out my 'Have a beer, cobber' routine, but instead I thanked him sincerely and jumped into the car. Home was Suffolk, and it was a very long way away, and anyway it seemed more kind to allow him the illusion that I was an overlander just starting off for Wagga-Wagga with our Williams lambs.

Instead, of course, the Williams Five came safely home to Dolgwili, small rather browny-speckled-faced sheep with just enough hint of the small Welsh Mountain sheep to cause us to fear, as soon as we got them on to our patch, that they would be flighty wanderers. In fact, once trained to our electric fencing, they settled beautifully, but they, too, were for ever as aware of their Williamsness as were the Suffolks of their Suffolkness. Whenever they had the chance they displayed an unostentatious cliquishness.

Golden Boy cared nothing whether they were Speckleface, Suffolk or Williams all through the September nights of the mating season, anointing them with the red or blue of his raddle to denote for us that he was fulfilling completely the demands of life. At the same time he achieved, however, indirectly, a greater integration of the various breeds within the flock, for by January all thirty-two females were gathered in from their free-run-of-the-farm to be paddocked for the rest of the winter in Jubilee Paddock, an area of no more than an acre, where the opportunities for cliques vanished.

Bounded on the east by a securely fenced and sheltering Morgan's Wood, on the south by some of Simon's best fencing and beyond that the river itself, on the west by the fence and hedge of the garden acre, and finally on the north by our movable units of electric fencing, the paddock was secure and cosy. It was easily accessible and within view from the garden. It sloped well to rid itself of winter's rains. It had the benefit of a small stream for drinking-water, but, above all, the overall concept of the paddock made us feel comfortable in our minds. Instead of the flock roaming off across the face of the land for almost non-existent food, breaking out and being who-knew-where, poaching up the gateways and crossroads of the farm, here they would stay, where we could see them, in touch and under our eye. We would bring their food to them, we would be able to catch them as they ate if they required treatment and as. they became due to lamb we could move them through on to Garden Paddock by day and into the shelter by night. The long and awful winter months at last would be bearable.

How we laughed and rubbed our hands as we pushed our spanking new sheep rack through and loaded it up with two bales of hay sweet enough, as we broke into the bales, to make me want to try a few mouthfuls myself. How we smiled and coo-ed as the barrel-shaped ewes gathered round and delicately selected wisps of hay through the mesh, and how warmly we approved of that splendid feed-tray below the hay rack for our concentrates. What a good winter we really were having. By God, we preened ourselves, what really good sheep-farmers we might yet become.

27   So much happens in a single day at lambing time

Feb. 2nd
At 3 p.m. Phyl called me to the wood and there we watched twin
lambs being born to No. 1 Suffolk. It was sunny and pleasant,
not cold at all, almost as if she had chosen her time. We stayed
all of an hour. I took photos, although I missed the actual birth
of the second lamb because the ewe was sitting atop of its
firstborn and I was wrestling to rescue it. The fierce slope of the
ground there, the dried leaves and broken boughs and twigs
made it a messy and risky birthplace. We took the family down
to the shelter later. The ram weighed 9–14 and the ewe 8–8.

Feb. 3rd
One lamb got chilled in drastic rain today. This morning the
ewe's udder was very lopsided and unnatural. I noticed the ram
lamb not seeming to get milk from it, so I pinned the ewe to the
wall and tried to strip the teat, but unsuccessfully. After the
chilling the lambs looked dismal and the teat still distended
unnaturally. This time I stripped most forcefully and at last
freed it. I held the ewe's head while I put a lamb on the teat.
Now the lamb enthusiastically filled its belly. I kept holding the
ewe's head at first but she stayed so still – in relief, I supposed –
with the lamb still sucking, that I let go. And still the lamb
sucked and sucked. Counting to myself, I estimated the lamb
sucked continuously for eight minutes when, becoming
alarmed, I removed her. As I left, the lamb had gone back for
more. Sad One has eye trouble.

Feb. 4th
Very bucked with my silage, unveiled today. The ewes showed
obvious preference for it and now, of course, I wish I had made

more. The polythene sacks in which I made it are now shown to be the wrong shape; I need a kitbag shape. Flock consumed almost two bales of hay yesterday. Ram lamb gained 1lb 3oz and the ewe lamb 1lb 8oz in twenty-four hours. Can I believe my scales?

Feb. 7th

Contagious ophthalmia spreading. Flame Gimmer 81 has it now. Treated today. Junglefowl bantam sitting on eight eggs in the hedge.

Feb. 11th

As I walked Caramelle after lunch I saw Darkie on her side straining. It was a prolapse of the cervix. Vet came straight out, we upended her and tied her to the gate for support whilst he replaced it and stitched her. The trouble is caused by her being so large with lamb. She has ten days to go and we must release the stitching when she starts lambing, nursing her in the shelter until then.

Feb. 12th

Was it only yesterday then that it all started? Now, 10 p.m. on Monday, it seems a week away. At 8–15 a.m. I found that the gigantic Suffolk No. 3 had given birth to twins in the coldest dawn we can remember since we've been here. The twins had somehow wriggled out of the shelter doorway and been unable to find their way back – the ewe was inside the shelter unable to get out to them – so that both were weak. When we had rescued them we found the ewe's udder looked wrong and the lambs unable to suck. The ewe lamb weighed 9lbs and the ram 7lbs. We brought them into the kitchen, warmed them by the Aga for two hours by which time they seemed strong enough to go back. The ewe lamb certainly had a long drink – I counted 170 – but all our worry was for the ram which at one time we thought was a 'gonner'. After two more hours in the kitchen however he was baaing his head off. By now we were milking the mother – once I drew off 7 oz – and bottle-feeding the lambs. We had two other worries; the hay had not arrived and Darkie was in distress. We went to bed very tired, leaving the stronger lamb snuggled up to the ewe in the shelter. It was not freezing out there but we reckoned it safer to keep the tup lamb – now saved by bottle-feeding – beside the radiator overnight. This morning however the stronger (ewe) lamb was dead and the world white with snow. The ram lamb was strong now but we could not get him to suck from his mother and I realised she had mastitis. By afternoon the day had become frantic. We had to rush to the vet for antibiotic treatment, Darkie had to be freed from her stitches, for now she could not pass water, we had to buy lamb-milk substitute and keep bottle-feeding the lamb, and then the lorryload of hay and straw arrived. Furthermore Suffolk No. 8 gave birth to twins this morning.

Feb. 14th
A glorious sunny day with all three Suffolks content on the lawn in melting snow. Trudi stripped the mastitis ewe whilst I had an extra hour in bed. We've named the bottle-fed lamb Homer and put him back with his mother for he was getting attached to us. He took 25 f.oz from the bottle today, some colostrum, rest substitute. The conservatory temperature climbed from freezing to 86 in three hours.

Feb. 15th
Homer is fine. Despite all-day frost he put on 10 oz. This ewe's udder is obviously healthier and Darkie is passing water satisfactorily now.

Feb. 16th
Udder still hard in places but Homer gained 1lb in a day. Checking at 10 p.m. I found Suffolk No. 1 in the middle shelter helpless on her back on the straw. Even after I'd rolled her on to her side she still could not get on to her feet. Her upsidedownness had caused her to swell and she took almost ten minutes to subside and become normal. Injected Dainty and Cripple against clostridial, and No. 3 again for mastitis, now almost gone. Marvellous treatment. Simon arrived with two dogs. He is one of 89 who did not get the rabbit-catcher job over in Lincolnshire. He is on holiday this week and still set on professional rabbiting. Darkie's bag is showing well; still twice the sheep she was last year, apart from the prolapse.

Feb. 18th

Sheep, it's all sheep. Our routine day changed at 4 p.m. when I noticed Sad One in a sheltered spot up the Steep. (She and Black Collar had been freed from Jubilee Paddock because they were unable to or unwilling to eat hay and had gone thin.) I went up. She had given birth to one lamb and the foot of its twin was showing. Phyl came up and stayed watching, whilst Trudi and I fetched six more bales of hay up on the kids' toboggan, for it snows and freezes all the time now. We 'sheltered' the Suffolks with their lambs and Darkie, and Trudi replaced Phyl in watching Sad One who still had not produced her second lamb. Dusk was coming, so I went up and fetched Sad One down to the shelter. She came quite unfussed and undistressed, although obviously it was a malpresentation. The vet was out, I was exhausted. Phyl phoned around the village until she got a message passed to Simon. Whilst I ate, she fetched him. The feet were still showing but we thought the lamb would be dead by now. *7 p.m.*: Simon soaped up and whilst Phyl and I held Sad One, he pushed the lamb back and righted it – twice. I admired the way he worked. Then, as she pushed, we lowered her rump and with Simon helping, out came the lamb alive and well, even strong.

Feb. 19th

Darkie ignored her prolapse stitches and delivered strong twins. Sad One started the day well, eating hay and nuts and taking water, but as the day developed she seemed uneasy. Tonight she is off her feed. No doubt she is sore from the lambing, but I feel anxious for her. Occasional snow, bitterly cold.

Feb. 20th

We are very lucky here again. Snow chaos all round but all is well here. Sad One poorly. Took her down to the vet to save the four quid they now charge for the journey. Miserable wait in the car until he arrived, held up by snow. He criticised us for not using an antibiotic on Sad One after Simon had lambed her, but thought she would recover after a three-day course of treatment.

Feb. 21st

Sad One was dead this morning. Her lambs are in Trudi's house being bottle-fed now. I was angry with myself at not knowing of the immediate need for antibiotic treatment, but my regrets for her passing were eased by the knowledge that she was old, had done well and died among friends in the total fulfilment of her function. We still marvel at our escape from the blizzards which have hit everybody else.

Feb. 22nd

A slogging day, this, slippery in thaw, with endless journeys among the flock, sorting out next to lamb, injecting with Covexin, feeding lambs, and moving those ewes with lambs. Dainty, Darkie and the Suffolks were happy on the front lawn. Nice Fleece next to lamb.

Feb. 26th

Ill. The last three or four days and nights are a feverish haze, with much sleeping and a sort of suspension of the time-sense.

March 1st

Up on watery legs but had to go back to bed.

March 2nd

Phyl has just come in to announce the birth – at 10.40 p.m. – of a single to Pink Gimmer. At 10 p.m. when Phyl was out bottle-feeding, the baaing of Pink Gimmer in labour had alarmed her and as usual P. had, I think, become over-concerned.

March 3rd

Bad night – nightmarish and feverish when I did sleep, full of coughing when I did not.

March 5th

Phyl has taken my cold. Black Collar lambed twins and another gimmer has a single. I went out a little today but I was useless.

March 6th

Buckets stood ready to be filled, dogs waited for exercise, logs to be moved indoors, all the sheep-work, my neglected seedlings, chore after chore shouted for attention. I turned my back on all of them except to go over and look at Black Collar who, Phyl reported, is going the same way as poor Sad One. Vet came out immediately, found her temperature was 106 and injected into a vein. She is better tonight but she has no milk and her twins are hungry. Poor soul, she has no teeth to keep herself in condition during gestation. Orange Gimmer, on the other hand, produced a twelve-pound lamb, our largest ever, and Nice Fleece bore a single without trouble too. That's what we want now, God; no trouble. Phyl has the worst cold of her life. Twice I watched her fall asleep by the fire with TV on, something I've never known before.

March 7th
Beautiful day. Daisy gave birth to a large tup lamb at dusk and there is a new tiny Suffolk lamb with legs so bent it can hardly walk.

March 8th
They've let us down on hay again. I grew angry on the phone. Little Maggot looks helpless.

March 10th
So much happens in a single day at lambing time that each day seems like a week. I whizzed Maggot down to the vet where she tottered about his shiny floor but he smiled and assured me immediately that 'she would come', that he had seen many worse than that and certainly not to have her put down. Certainly she has gained weight despite her crookedness. Went out at 10.30 to check on Wild One and found that Red Balloon had just had a lamb. What to do? She is nervous and the lamb was still wet. We decided to leave her an hour. At 11.20 I reported back to P. that she was okay with the single and we reckoned it best to leave her out for the night.

March 11th
I awoke at 6 a.m. and through binoculars at the bedroom window saw that Red Balloon after all had had twins. Later Wild One had twins too. The grandkids have named Sad One's orphans unaccountably Jessie and Robin. They're now taking

228

four pints a day between them. The lambed half of the flock are getting short of feed and pushing down the railway and along the river. We're pushed for cash too, although I was able to pay for the hay which arrived today.

March 12th
Let Wild One and Red Balloon out into the sunshiny day and they grazed all day up on the Steep. By 6 p.m. I was almost out on my feet despite a doze in the chair from 2 to 3. Phyl is worse but she will not rest and just mooches on, just like C.K.

March 13th
Filthy, depressing day. You don't walk; you skate on mud. Blue Gimmer had a single in the rain. We put her in shelter with WO and RB and then seeing Spotty close to lambing put her in too.

March 14th
Spotty had twin lambs when I went over at 10 p.m. tonight.

March 16th
My first sight of the world this morning was shocking; the garden was covered in quite deep snow. Later however it became a nice day with definitely a sight of new grass. We let the latest lambings out and late tonight Whitey gave birth to a large ewe lamb. I have the impression that the Suffolks are no longer suckling their lambs much.

March 17th
We were going about our sheep business normally this morning, setting up the scales on the sawn-off bough of the elm to weigh Whitey's new lamb when I was quite overcome with sudden weakness. Suddenly, en plein air, all I wanted was bed and sleep. I charged indoors, somehow got myself a coffee and clambered into bed, clothed. It was 11.30 and I slept solidly until 5 p.m. Even now, after a meal and no work, I feel fagged out. Reckon I got Farmer's Lung, mate. While I slept, Little Gentle had another ewe lamb but by dusk had done very little mothering of it.

March 18th
To hell with everything.

March 19th

After 5 p.m. feeding and 'doing up', Phyl found Spotty with only one lamb, crying and crying for her other one. She went repeatedly to the bottom of Garden Paddock to call down towards the river for it, so all of us formed a search party. It was eventually 10.20 p.m. when the three of us gave up after systematic listening-out sessions all along the flooded river. Several times we thought we heard a lamb cry against the background of the river-roar but nobody was able to pinpoint it. Although there are dangerous places down there where the bank is undercut or gullies enter the river near where two large trees have been uprooted, I found it hard to believe that the lamb had drowned. We went sadly and very tiredly to bed.

March 20th

I did not wake till 9 a.m. and even then fell asleep again. I have an idea I'm somewhere near the end of my tether. Late in the afternoon a miracle happened; Spotty's lost lamb turned up suddenly by itself in Shelter Paddock. It was a bit feeble, but quite dry, walking, and not at all bad considering it had had nothing to drink for twenty-four hours. Spotty, however, refused it utterly and when we shut them up together would have killed it so rough did she become, charging it against the concrete wall time after time. So now we're bottle-feeding him – Survivor, he's now called – together with Jessie and Robin. Today Phyl freed the whole flock from Garden Paddock on to Sheepdip. The ground is quite sodden.

March 24th

Bright, windy, drying. The sheep are scattered all over the place, including eight through on to Thomas'. Good Friday is neither fish nor fowl, not a holy day any more, not a holiday; nobody seems to know quite what to do. Today we Covexined the remaining eight expectants. Four Williams are in lamb, Phyl thinks. This morning we began tidying the garden a bit, gathering up the wasted hay, scratching at all the sheep muck left on the drive, emptying the trailer. I felt lighter-hearted as wind dried the earth. The feeling grows that winter really cannot last that much longer. Even the shed door has dried enough to close, that all in the last two days. Thomas' nephew came by and confessed that they have had 'very heavy losses' of lambs up there due to the severe weather. We still regret having lost our single lamb early on, but evidently we have come out of it all fairly well; nay, very well. I really must try to be more grateful for all our blessings.

28 Us sheep-farmers

"Where's Sam?" Deenah astonishingly asked as soon as she arrived inside the kitchen with her father on the usual Sunday visit.

"Hullo, Deenah," we chorused in greeting, smiling with pleasure from our chairs.

"Where's Sam?" she demanded again, her tiny face red from the cold outside and her eyes smilingly, sparklingly cheeky. She was bundled up in layers of multi-coloured clothes and wore bright red wellingtons.

"Fancy her noticing that straightaway," Phyl muttered at Simon as she went forward to pick her grand-daughter up.

"Where's Sam?" the child asked again.

"Poor old Sam died. He was very old, older even than Poppa. He wasn't very well. He's asleep up in the orchard."

"Under the James Grieve," I assured her, lest she have qualms that we had not done our best for the old dog.

I had not wept for Sam-Boxer, for I had not watched him die. Not seeing him die, but only dead, I could tell myself that he had had a long and good life, that near the end there could have been little pleasure in those rheumaticky old bones and those lustreless eyes. He would have collapsed instantly at the prick of the vet's needle. Later I had seen him comfortably installed in the large deep grave I had dug for him the previous day up in the sloping orchard. From five such graves now the ghosts of our dogs could watch the slower comings and goings of those who had shared their days. Tiger Lil and her son Urchin were buried side by side down near the farm gate in Garden Paddock, Mini the poodle was in the garden, the boxers Sam and Jemima were in the orchards. It was the absurd sort of notion we harboured about our animals, a middle way between the animal-as-brute and the animal-as-human approaches which seemed to show I was more and more attracted to the Chinese Taoist way of seeking harmony and grace, what little I knew of it.

"You want to get a proper dog now," Simon said eventually. Phyl had just boasted to him that we had at last ordered our sheepdip, and had been on the point of suggesting that he dig the hole to receive it when he jumped in with this sheepdog notion as if finally to complete our sheepfarm concept. "My boss has got a sheepdog puppy spare," he smiled challengingly as he endlessly stirred his coffee.

He had pricked us more than he knew. We sat at leisure in the Sunday morning kitchen, three generations considering dogs, for Deenah had finished her blackcurrant syrup and was bopping impudently to fingerpoint Caramelle's silken ear and to peer into her almond-shaped eyes.

Sam and almost all our dogs had failed to be proper dogs in Simon's opinion because they had no function but that of companion. Proper dogs hunted, or guarded, or shepherded, or guided. I nearly began to argue a case on behalf of Sam, who having being bred for the showring but failed because it was not in him to swank and brag, became instead the most lovable and reliably friendly of companions. I did not so argue, however, for already that sheepdog puppy was leaping about in my cautious mind.

Hills had grown steadily steeper, our steps slower, our tiredness greater. We coughed regularly. Little bits and pieces of health fell away from both of us in the aftermath of last lambing. Always there was so much work, never were the days long enough or body quite lithe enough. Algae and spiders flourished in corners of the conservatory, border re-plantings and seed-sowing postponed themselves, care languished. We sat down more often, picked things up less frequently, and harboured in our minds yearnings for a small personal helicopter, a small Dorset farm without hills, a windfall, wings, anything to make work easier and more productive.

We had come to admit, not yet that our days were numbered, but that they were lumbered – with age. A time must come when we really could not climb the Steep or capture an ewe; towards that time we really must get things right for ourselves.

233

We needed to reduce the size of the flock, we would tell each other.

"Save you no end of work," Simon smiled. I smiled too; did our son really not know that the real reason he urged a sheepdog on us was so that he could train it?

"We've thought of that more than once," I told him.

"Could you train it?" his mother asked, with an edge of doubt. Too often, I'm afraid, that edge of doubt showed, hers and mine, when we considered his abilities.

"Yes."

How? When? I wondered.

"It's working now a bit, in the yard, with cattle."

"How old is it?"

"About three months."

"How much?"

"Not much."

"Where's Sam, Poppa?"

"Up in the orchard, sleeping."

"Poppa, where's Sam?"

"Heh, watch it, mate."

"Where's Sam?"

The damned puppy wouldn't lie still. I hugged our grand-child.

"We'll think about it. Will it still be available in a month's time? End of August, say?"

He nodded, happy enough to have got so far.

The dream of having a really good sheepdog to work for us had always been and still was as tantalisingly out-of-reach as once, long ago, Phyl Kirby herself, pilot's wings, a two-acre garden and a host of other wonderful things had seemed. So utterly desirable they appeared, that really they could not be for me.

For one thing I had seen enough of sheepdogs which were less than 'really good' to know that such dogs made work rather than eased it. I had seen poor Thomas himself running up and down our Steep to catch one of his strayed lambs – a fearsome sight in itself – and made more critically breathless by his angry shouting at his idle sheepdog than by his running. I recalled too the frustration of the two shearers who had brought their sheepdog to gather our flock for shearing once. Our donkey started a chase-me game with the strange sheepdog, which became so startled out of its wits at the sight of Marigold's repeated charges with head lowered and legs kicking and an occasional extra loud hee-haw that the dog ran off down the railway bank and refused to heed a word or whistle of command all day long. I knew too that most of the sheep-worrying in the district was caused by out-of-control sheepdogs. Warnings in the classified advertisements of the local paper were inserted by farmers weekly that any dogs found trespassing on their farms would be shot. At the same time, none of all that had totally dispersed the dream of being 'One Man and His Dog'.

There was Caramelle to consider, of course. However differently she behaved now that Sam did not sleep at her side – she had freshly developed a little melodious howl of greeting for us, as though relieved that we at least had not deserted her – she was hardly likely to lavish much affection on an upstart sheepdog puppy suddenly invading her long-established territory.

Yet how ever could we consider ourselves proper farmers without a sheepdog? And wouldn't it really be a joy if this one did turn out to be a good one like Juno? And didn't we just need such a new interest at our age?

At our age, caution was first nature. All through the sheep-easy summer months when farmhouse guests took over our lives we could fully indulge our caution about taking on a new puppy and about taking on a new breed of sheep, for both decisions could be postponed at least till the end of August.

The realisation had come to us that really we had a very good set-up for breeding pure-bred sheep, yet we did not have the ideal breed. The Beulah Speckleface really was a splendid sheep and we admired their hardiness without reservation. Too many of our Speckles, however, had led us too much of a dance through the countryside these last few years for us now to be willing to dedicate our lives to them. We wanted something less of a wanderer. Well, the Suffolk did not wander; perhaps it did not wander enough. Nevertheless – and it hurt a little to admit it – the Suffolk and we simply were not a success together. The original three ewes had always borne good lambs, it was true, usually twins of good weight which had looked beautiful and prospered exceedingly at first when we had still been able to weigh them on our spring scales, which had a limit of forty pounds anyway. After that something always went wrong, something we could never define. The lambs would go thin and miserable, often with scouring. They granted us no pride in them. We had run Plato and Hercules on to become yearling rams and certainly they did gradually improve with sunshine on their backs, but in our hearts we knew they were nothing compared to the magnificent young rams we always saw at the ram sales. Ours had no real quality. The Suffolk ewe lambs that we kept always seemed to do better, but even they were not what we had hoped for. Suffolks, we came to admit, were a disappointment. We had wormed, we had treated for cobalt deficiency, for fluke, we had blood samples taken and consequently treated for copper deficiency thereby proven, we had worried and consulted and read up, but at the end of it all we could only shake our heads and say we had failed. They were, in any case, the most popular of pure-breeds and around us were old-established flocks whose flockmasters probably had generations of experience in producing marvellous animals for the autumn sales. We simply could never hope to catch up on them.

True the Suffolk-Speckleface gimmers promised very well indeed, but what we wanted was a pure flock, not crosses. Our flock by now was uneven in age and breed. How much nicer an even flock of young Hampshire Downs, say, would look. Or something a bit unusual; Jacobs, or Wensleydales, Devon Longwools . . .

"How about those black sheep?" Phyl would say in the middle of the night.

"How about Texels?" I would counter at breakfast.

Texels were winning every competition in sight, for they had the capacity of putting on lean meat without fat, the clever ones said, and this was suddenly so infinitely desirable and necessary that demonstrations were being given regularly to show us most impressively the old bad ways of fat fat lamb contrasting with the new good ways of lean fat lamb.

"It's got to be Texels," I would assure her at lunch.

Summer to us sheep-farmers is rather like winter to us gardeners, a close season, a time for dreaming of all the glorious sheep we will grow, all the lovely plants we will rear, for browsing through the literature put out by sheep-breed societies and seedsmen. There is no hurry. One can speculate endlessly, change one's mind nightly, plan and re-plan, pent and repent until that moment arrives when the axe of decision must fall to kill all dreams stone-dead.

"Well?"

"Certainly not Texels, anyway. They've gone mad. Four thousand pounds for a ram lamb somebody paid at that sale you were so keen to go to. Thousands of pounds. Somebody's made a killing. No. I fancy Llanwenogs. I think they're just right for us. Local. Not too big. The only trouble is we've missed their Society Show and Sale, I believe."

"Oh dear. What about the puppy then?"

"Oh, I think so, don't you?" She laughed, her eyes shining wistfully.

At our age caution could also slip out of the window.

"I must admit it would be damned nice to have a healthy young dog, well-trained, about the place again."

I was remembering poor old Sam-Boxer. This last year he had been a mess of small sores on his feet, of irregular patches where he scorched the hair off his body as he tried to get even nearer and even into the Aga's heat to warm his old bones. He had begun to smell old-doggish, his eyes grew sunken and his step slow and uneager. All his comfortable life he had been most accomplished at making ghastly noises, a sort of party trick, snoring, licking himself, eating, breathing or just plain snorting for the hell of it, and in this last year, whilst other physical processes degenerated, Sam's skill at making revolting noises improved all the time. Caramelle had aged too. Grey had begun to drab that glorious chocolate, gold and white of her long pointed face, and worst of all, her coat had become a thick rug of lustreless hair which no amount of grooming could improve for long. Since her escapade with the lamb, her opportunities for running had had to be curtailed. Five years or so ago, Caramelle and Sam had run free all over the hillside, unfettered by sheep, marvellous in motion. I could remember exactly the sound of Sam's heavy breathing as he lumbered down the hill in the dusk, unseen but so plainly heard, whilst Caramelle pushed her head against my hand as we waited for him. Curtailed by sheep and age, we were, all of us.

"Yes, wouldn't it be nice. We'll tell Simon 'yes' then, shall we?"

Bess was a beauty right from the start. I had almost forgotten that a dog could be so lovely, so taken over by old dogs had my life become. She glowed with health. Her splendid eyes were keen and lively, and as if the sheen upon her coat and her well-balanced Border Collie markings of black and white were not beauty enough, Bess was further blessed with flashes of pure chestnut on her face and legs. Had she been boss-eyed, had only three legs or barked all through the night, some doubts might have remained in my timid mind about taking on a new dog so late in the day, but her beauty dissipated all doubts. She must stay.

Caramelle might kill her, of course, as Simon suggested. Or Bess might kill me, for already she showed she had enough energy to run us both easily into an early grave. And she was almost completely untrained with sheep, of course. Otherwise we welcomed her with open arms, which of course could not contain her, so energetic and excited was she.

We saw no point in putting Simon's drastic opinion to the test, so brought Caramelle from her shed to lord it, or lady it, indoors and gave Bess the shed. When gradually they began to meet, each on a leash, they already knew of each other's near-by existence from smell and sound, but, in fact, at the first carefully-arranged confrontation, the tall and aristocratic saluki it was which affrontedly backed away from the eager advances of the prancing puppy. My total belief in the power of territorial imperative assured me of the wisdom of allowing the two dogs to get closer acquainted only on neutral ground. We therefore hoisted Caramelle into the back of the Renault, and tried to contain the overflowing Bess on Phyl's lap, whilst we drove to the beach. There Bess treated Caramelle as though she were just one more dog, although Caramelle, quite unable to be that insensitive in return, quivered continuously, curled a lip in warning, and took to looking out to the far sea-horizon rather in the manner of her master and mistress when confronted, say, by some weird extremism from the younger generation. We kept them on leads for a while, but when unleashed Bess made bouncing approaches to the saluki as tiresome as those of a grandchild to a tired grandfather, yet triggering no more violent reaction. The collie took instead happily to running about fifty-seven miles across and all over the pastel acres of that deserted, Saturday-afternoon beach – for that day our bit of Wales was trouncing all-England again at Cardiff Arms Park – whilst Caramelle walked sedately beside us on her lead with never a show of real passion but with rather an air of simply not knowing what things were coming to – again rather like ourselves in a wider context.

Days later, I met a new stroller along the old railway track, who, seeing Bess and being apparently a retired farmer, paused to chat awhile of sheepdogs. Welsh farmers, he told me, used often to train their dogs on ducks in the farmyard for starters. It was easy and convenient, obviously, but being a bit short on ducks, I was unable to try this – or another, of tying the puppy to an older working dog – as I was unwilling to try a different method, that of strapping the front leg of an impetuous puppy inside its own collar. Those advices, and the one we already knew, of taking the puppy among the flock on a lengthened lead for training, left me a bit short of what practically I must do next to further Bess' education, but fortunately I could shelve that for the time being, since real sheep-work is not usually started at under a year.

"Is she still interested in sheep?" Simon would ask on his weekend visits, for sometimes, it seemed, even sheepdogs went off sheep.

"That, at least, is no problem," we could always assure him.

She had but to glimpse sheep the other side of the hedge and she would be away from lawn-training to wriggle through hedge and fence – she was still small – to chase them uphill, positively oozing intense sheep-interest. The donkey, perhaps remembering the fun and games she had once had with the shearers' collie, usually joined this merriment by chasing Bess whilst hee-hawing her great head off, whilst a worried Phyl chased both at her intermittent jog-trot, screaming for Bess to come back here at once, you bad dog, and I brought up the rear at my imitation gallop, shouting angrily in my reedy voice until it and breath gave out and we all gave up, to stand there in silent breathless frustration, at which the dog invariably returned to us in expectation of praise and biscuits. As yet, obviously, she was not safe to be free in the garden.

A working dog, everybody said, will eat up a hundred miles a day – a fact, if indeed it is, which immediately overwhelmed me with despondent tiredness. A puppy, surely, must need twenty miles a day to keep him fit and happy? Why on earth had I ever been so foolish? Anyone would have thought that by now I would have achieved some modicum of sense to keep me from all this sheep and sheepdog nonsense. All this had never been the idea at all.

So we tramped with her on a lead across the farm, and we tramped with her for miles along the old railway track off the lead, and we played with her on the lawn and trained her so that she may have had all of six miles a day. Perforce she was shut up for much of the time in the shed where she exercised herself chewing off all the knobs from the drawers of tools, eating half a table-leg and two bags of kindling and a shelf-full of garden requisites like lime and quassia, before starting in on Phyl's fleeces then awaiting spinning. Finally she perfected a technique of mining down into the admittedly ill-made concrete floor. All this damage she managed to achieve in the very nicest and best-tempered way, smiling up at us when we opened the shed door, never barking to be allowed out, never bolting and never messing. Simon, we told each other, had picked us a winner. If only she and Caramelle could be trusted to play and exercise themselves together in the garden, Bess' coming to Dolgwili would be a most happy affair.

Of course, it was Caramelle's garden. To meet the upstart puppy on the beach was one thing, but here on her home patch, that was quite another. This the saluki imparted to the frisking Bess with all the authority of threat she could display. Bess was still only a quarter the size of Caramelle, and we were timid people. We bought a muzzle.

Part of my war-hangover still was, not expectation of disaster, not intimations of disaster, but a sort of narrow-eyed lookout for it. Tragedy might still be near. A strange man on the stairway, fox, a drowned child. Careful, the hedgerows warned. Watch out now, the daily circumstance whispered.

So we muzzled Caramelle. I resented it as much as she did. The muzzle was a gawky and undignified contraption of new leather which never fitted her comfortably and which she wore with ill-grace. It granted us some temporary relief from that worry of having those long jaws savage the young dog, but gradually our dislike joined with the saluki's dislike of the muzzle and we took it off. For all her early barking and lip-curling, the old dog realised she had to accept this new companion and although she never joined Bess' frolics in the true spirit, day by day she was learning to live with this new order of things. Yet even then, this Caramelle-Bess exercise was a prickly matter in our minds; Bess might yet get pinned down and badly hurt, or both might take off if, for example, we had misjudged the movement or whereabouts of the flock. In theory this whole garden area was safely fenced or hedged and gated, but in practice there was no such thing. The goats were able to prove this fairly regularly, and if goats could get in, dogs could get out. Anyway, the exercise brought no real ease of mind, and whenever the two dogs were out there together slightly uneasy eyes were always watching them from the terrace or the windows.

Nevertheless, Dolgwili did now have its own sheepdog, however young and untrained and poorly exercised. The self-mockery with which we had long been accustomed to refer to ourselves as 'us sheep-farmers' would no longer sound quite as mocking or seem quite as deserved now that we had Bess at our heels. Indeed, when Simon from time to time reported back to us some of the ignorance, mismanagement and even cruelty he had come across on other sheep-farms, we realised that we must by now have hauled ourselves at least halfway up the league-table of sheep-farmers. We had learnt a great deal, we had come a long way. We would always be small, always suffer from Dolgwili's natural deficiencies, but once we had our sheepdip, once we had decided on our permanent breed of sheep, and once we had persuaded Bess really to save our poor worn-out legs, well, by then 'us sheep-farmers' would probably be unwilling to endure our own self-mockery at all.

29 The morning after the battle

November 2nd
Disaster.

It is 1.30 a.m. as I write. We are both shattered. Phyl is in the kitchen, unable to sleep, pottering about and I don't know what else I can do. Sometimes I go outside and stand listening. I was fishing this morning – not very seriously. Lately I've been going down there often. Just to mull things over. It is so private. And perhaps the everlasting river-sound flushes out these anxieties. Anyway I took the rod down to this favourite place, the secluded beach of shingle under the tree roots which nobody else ever fishes and where twice I've seen kingfishers. I sat there thinking what a damned lottery everything is. I was enclosed in natural peace, yet I was never at ease. The sheep kept drifting into my head. Once or twice I thought I heard the dogs bark – Bess and Caramelle had been out on the lawn when I left – but down there in the chasm it is impossible to know where sounds come from, and I was wearing my hood up and that makes rustling noises against my ear. Then, beyond doubt, I heard a foxhound giving tongue. Almost simultaneously three hounds came rushing down the bank opposite, straight across the river and then splashing diagonally up and heading for Jubilee Paddock. Christ, I thought. I clambered up from the river in a sort of dark despair. I could hear the hounds baying as I got near the top and another half dozen streamed past me as I got into the meadow. Quite obviously they were on to a fox. I could see them scrambling through the hedge and over the fence into Morgan's Wood. I looked up to the lawn. Trudi was out with Phyl near the house. Of Bess and Caramelle there was no sign. I didn't have to ask; their faces told me. How long, I asked. Minutes. I fetched the binoculars, but even then I think I knew that the saluki was as good as dead. Bess too, maybe. It wasn't that; it was the sheep, everybody's sheep. When I focused on the wood I could see the hounds climbing, going away through the trees. I did not see our two but once I saw a brown dog that looked like a lurcher. Simon and Caleb had gone along the railway, Trudi said. So. We stood about, saying foolish, unnecessary things. So far there had been no sign of the hunt but we began to hear a horn from along the railway. I walked down. Two riders came down the track. One wore hunting pink and rode a white horse. With him was a young woman rider. He sounded the horn again and came up to the gate. He seemed to think I was about to open the gate for him. I held it shut. I was angry and I was distressed. He was young, very healthy looking, smiling. He said, "Have you seen any hounds?" at the same time as I said, "By God, you've made a right mess of this one, haven't you?" I told him what had happened. It had started to rain heavily. I still did not see other riders. Just these two. It seemed such a crazy hunt. I suppose I got a bit incoherent and he took to sounding the horn some more, looking across to the wood, wondering what to do. Really, even now, hours later, all this seems such a screwing up of circumstance, just like the night God Himself got shot down in flames when I went to the tail for a piss and the bit of flak came right through my seat and killed the second dick instead of me. Crazy. Everything is such a damned lottery. This proper hunt only comes this way once in several years and it must be a pretty hopeless exercise anyway. Then for them to come out today, and to come on us in that way, without chance of warning. As the white horse grew restless, the young huntsman must have realised without my repeating that he could not possibly ride through that wood. I don't know anything about hunts but it seemed to me that his pack were gone to Kingdom Come. He asked a few questions and the girl began to worry a little about the rain beating down on to her elegance, to suggest they try down there. Why do I write these useless things? He remained very civil. I told him about Caramelle. I pointed out he could get on to the road near the bridge or try down the railway track and after a mumbled discussion they ambled off towards the bridge. He had said something about the hunt secretary would come and see me about any damage and I said something about . . . I don't know. It was

243

such a shambles. They stood considering what to do down on the bridge and he sounded off again. Two hounds did appear a bit later. The huntsman slid off his magnificent horse, opened the gate for his companion, and they trotted off down the road. I stood about in the rain, searching through smeary binoculars, hoping for the return of a dog I had once known. Another hound made a detour of the Steep, where our flock had been standing defensively nearly all the time, staring out in the usual alarmed fashion at anything new. After a time I came indoors and phoned around. Neighbours, the police. None of us could settle. After about half an hour the two riders came back to the bridge. They had collected about eight dogs by now and he blew his horn a lot, kept looking across to the wood. It was pouring now. I thought that might deter Caramelle and send her home. I watched from the conservatory, with the daylight going. Even then I never did see another rider. I was thinking of the chaos going on out there somewhere. It was a misery. A while later, Simon-Caleb came along the track. Marvellous; Caleb was carrying a dead fox. They had got the springer and two Jack Russells with them, but both lurchers had gone. They had been working a foxhole on a bank high above the railway when the hunt came through. They did not say much. I told them our story and Simon said they'd drive around the back road, see if they could find anything out. As darkness fell, the hunt Land-Rover came through our gate on to the bridge and drove up the track. I could hear the horn being blown every now and again. It was still raining. Once we heard a hound somewhere round the back of the house and I had seen one earlier galloping back towards Thomas'. Even quite late I could hear one baying somewhere up the Steep. Simon phoned about ten o'clock but really it did not amount to anything. One lurcher had found its own way home. Simon was in the pub then and everybody was talking about it. I saw the lights of the Land-Rover go through our gate and off about half-past ten, presumably with all their hounds. So far we've heard nothing from anyone.

November 3rd
I went out this morning

The phone-bell had rung some time in the night or early morning or perhaps it hadn't. Dream and waking were the same. All the time I could not rid my mind of the rooky young doctor at Ford. We had fallen asleep eventually but I slept with a waking mind and it went over and over the same, obsessed with him standing there in his bright new uniform with flt-lt stripes and his new brown kid gloves in his hand. The VR in his lapel shone out and he would keep slapping his cupped left hand with the new gloves, and us still in flying clothes from the night. Steve, Tom, Lew, me, Rube. And this doctor still in his napkins telling us we ought to have saved the second dicky, Steve angry like I didn't know he could be and all of us raving back at the doctor, had he any idea what it was like up there under attack, the darkness, the flak coming up, the kite weaving, diving, throwing all over the sky, nobody knowing. All that arteroid blood next morning congealing in the cockpit floor. All the time, even when I went out, this silly young doctor tortured my mind. Several old ones, Darky, Dainty, Nice Fleece and poor old Daisy were dead already, their legs stuck stiffly in the air, disembowelled. Torn and chewed up. It was an abomination. You could only close your eyes and keep shaking your head. Others were quite shattered. One stood but on a leg which had been chewed by the dogs, nothing was left of the leg but sinew and bone. Entrails hung out, throats had been torn, eaten alive by the dogs. Perhaps that was why; the morning after the battle all over again. He had been a complete stranger, I had not even known his name. The dead and dying, the stunned survivors. And I should have been dead. Crows and magpies already picking out the staring eyes. I walked about among the carcases on Sheepdip. Spotty and Cripple came rubbing up to me for nuts just as if nothing had happened. People don't know. You can't tell them. It seemed so silent again, just like that morning, the great old Halifax gaunt and stone-cold, the blood congealing. Mulligan came over. It had been him phoning. He had seen the dogs, early. Yes, five and he thought our tall chocolate dog was one. It, three sheepdogs and a big black alsatian. Simon phoned and I told him. It took me a long time, but in the middle of Sheepdip was this great poached-up muddy part, circular, and I realised then that was where the dogs had been working the sheep round and round. Before they attacked. P. insisted on coming out. We stood about, huddling into ourselves, not saying anything. Sort of

245

despair. It always ends in tears, Lizzie used to say. Everything. Can you think of anything that doesn't end in tears? We were still waiting for the vet when Simon arrived. First thing he said was, "They're up there now," and when we looked up the Steep there were two dogs, a sheepdog and the alsatian. He ran down and fetched his twelve-bore from the van and by then there were six dogs. One was Caramelle. They must have come out of our wood. Wild ones, Simon said straightaway and we began to walk up there towards them. I saw P. walk away and I said to him, "You don't mind, do you?" He had been a schoolboy then and he had paid half the thirty-five quid for her. Nor did I now. It was over. She had let me down, and I had let her down. I called her once or twice as we climbed. It looked as if they would all get back into the wood. Simon's lurcher wasn't there, nor Bess, he said. Just when Caramelle started to come to me, three men with guns appeared over the skyline of the top hedge. The dogs had started to bolt. The guns fired in salvo. One by one every dog fell. I stood there. Simon went up to talk to them. Somebody finished a dog off. I wept gently. Mulligan said he'd get the boys over to clear up the mess. The vet had to put down seven more and Mulligan had shut a few wounded ones in the shelter. Six farms had lost sheep. One of the guns was Tom Davies but they didn't know about Bess or the other lurcher. Or they said they didn't.

I shall bury her alongside old Sam-Boxer. I brought Spotty and Cripple through on to the lawn and I left the gate open purposely. The donkey came in too and it did not seem to matter. I think I've just about had enough.

30 Epilogue

The grass and thistles had grown jungle-high all across Sheep-dip. Wilderness had set in. By late summer thistledown floated in clouds on the west wind all across the kempt garden. Occasionally a few retired old sheep, survivors, padded purposelessly down a regular path they had made through the long grass to sit it out contentedly in the shade of the shelter, eyes closed, dreaming like lost old ladies in an old people's home. They had not been shorn this year; their fleeces peeled from them in shreds, which Phyl collected and pulled from them for her spinning. They rested on solid old straw litter which had not been forked for a year, and through high summer swallows had swooped without hindrance into shady nests on the rafters of the shelter. Redshank, plantain, nettles, mayweed and thistle rioted over the untrodden pathways and posts nearby, and over the brows of Sheepdip and all over the Steep, forests of bracken prospered more thick and more tall than ever. Rampant hedges scraped the sky. Everywhere the face of Dolgwili had grown as bearded as my own.

No lambs had been born at Dolgwili this year. No sheep broke its bounds, no dogs or ponies chased across its forsaken acres. A solitary donkey brayed complainingly each morning and even occasionally at midnight, grew slowly eccentric and worse tempered, and chased the old sheep when boredom became unbearable. The garden-invading goats had gone, and all our fields and hedges were a disgrace, an area of uncaring. Hawthorn, bracken and bramble had won; they waved their long arms in wild victory, whilst we turned our backs to tend once more our immaculate turf and weedless island beds and rockeries. Everything in the garden was lovely; everything outside did not matter. Adam delved; Eve spun.

The period of shock and low spirits which had followed the Night of the Dogs had gradually eased, healed by time and perspective to become, quite incidentally, a period of holiday. What had seemed to be one of the arts of living – to find the glow of advantage within the hurt of disadvantage, to convert failure into success – we now achieved without really trying. Even as we turned away from the horror of massacre to the benison of the conservatory, turned our backs on sheep and bent them to the garden, our days and spirits grew lighter. New plants, new trees came. Garden Paddock returned to the garden. Time, endless time, flooded in. Spring evenings came gentle and unpressed by lambing. Nights of storm brought no sleep-interrupting fears for ewes, and the horticultural days slid slow and easy without a single urgent call by sheep. Morning sour-temper could be assuaged by morning in bed, days stretched before us almost infinitely long. We could fish, we could watch the woodpecker, we could walk the coastline, we could spin and knit and transplant seedlings, we could sit in the sun. We were on holiday; Dolgwili was at rest, after sheep.

But then, one lovely day in early September, Simon drove over in his boss' Land-Rover. We had been sitting outside, still right in the eye of the sun, discussing where we would go and live next, for the money had run out, and it was absurd to live on a place like this. The Land-Rover halted mightily under the cypresses near the house, and as Simon got out of the driving seat, a sheepdog jumped out of the back. Black and white, with flashes of chestnut.

"Come here, Bess," he called sternly as the dog began to scamper about the lawn. It came.

"It's Bess," Phyl said, standing up. "Bess, come on, Bess. Come on then."

They fell into each other's arms and I could hear an angel choir over Mulligan's way whilst we all laughed and grew suppressedly excited, and the dog went from one to another like crazy.

"Where did you find her, for goodness sake?"

"Llandeilo way. Boss sold a bloke a tractor and I had to deliver it. Recognised her in the yard. Been working."

"Working sheep? Really?"

"Bess, Bess, well, well, well. Get down now. Good girl. You *are* a beauty, aren't you?"

She was. Her coat shone, she was alert, with a coiled spring tight inside her, and so happy.

"He said she'd been working with an older dog all that summer."

The dog-happiness went on and on. Stroking, ear-rubbing, she holding our hands in her teeth, rubbing into us with her gorgeous head.

"But how did he get her?"

"Said she just turned up one day."

"And she's been no trouble there? They say once they've got the taste . . . Did you ever hear anything about the other lurcher?"

"No."

"Did you have to pay him?"

"I told him what happened. It's all right."

"Well, well . . . "

She looked no different. Bigger. Quite beautiful. She laughed a lot with her tongue out, and turned on her back in the terrace dust whilst Phyl played with her and fondled her.

"Are you going to keep her?" I asked Simon when we had made a pot of tea and sat outside in the sunshine drinking.

"Do you want her back?"

"I'd like her," Phyl answered immediately. "Yes."

After a while, slightly unsettled, I left mother and son and dog sitting there, to stroll without purpose across to Sheepdip. We had never found time to whitewash the exterior of the shelter and the summer had bleached the creosoted stable door and the timber cladding to grey. The sheepdip which had cost us so much, which had never been used and which had taken me all those cursed days of wearisome labour to instal level, now was half full of greenish rainwater. A drowned cock-chaffinch floated in it; nettles grew up into the protective surround fencing. The footpath, the sheep rack, cradle and troughs stood about with all the sadness of desertion. Little rags of wool still clung to posts where once the sheep of Dolgwili had rubbed.

It was a bit crazy to go. He had never actually said so, but he had somehow left me with the impression that he would not mind one day inheriting all this. I had always known that one

249

day a man would cut down my garden trees, would open the gates and let the sheep back across my garden; maybe it would be our own son, that man. It was hard to imagine that these hills would not always be here and that they would not always carry sheep. Besides, as long as I died back in Suffolk, or they scattered my ashes from the bridge at Brundon Mill in the shallows where once I angled the silver dace, I could bear living in this hard and lovely place a few more years. We would have to sell the grandfather clock, or the car. I knew the ways of the place now, and I had left my mark. The garden was a harmony, the conservatory sweet refuge, the home pleasant and cosy; yet Dolgwili was a place for sheep, made no sense without the on-going and ever-more story of sheep.

"I noticed the Llanwenog Show and Sale is announced again. 15th September," I said subsequently, slightly absently, a seemingly passing utterance without significance, but I could feel the wind of her sharp glance across my face as it passed.

"So?" she challenged.

"Well," I shrugged. The holiday was almost over; already my mind was pressed by the inevitability of sheep again.

The eleven tup lambs on parade looked as smart as a paint advertisement on television. Their extra black faces, made up for this performance, I did not doubt, were crowned at the forehead by a shapely crest of wool, and their whiter than usual fleeces had been lovingly groomed and shaped. As each young ram was inexpertly moved around by its handler – each sheep was haltered – in an approximate effort to display its good points, they pattered about the clean concrete in the sunlight on shapely black hooves and chic black legs which matched their faces and which made them seem to our sheep-starved eyes quite delightful. They were slightly nervous and sharp-eyed under the scrutiny of the ring of knowing spectators into which Phyl now led me by the hand, a little excitedly, a little quick with eagerness, to watch the judging.

The last time we had been here at this market eleven thousand sheep had been penned for sale, and the whole market complex was ajostle with business, with farmers, with traffic, a gigantic throbbing affair. Today was utterly different. It was small, friendly and almost private, and as at last the intent judge relaxed his gaze to indicate his decision on which ram was that much better than the other ten, we joined in the small, rather polite, round of clapping, we smiled at each other and remarked how nice this all was, how it was our sort of set-up, our sort of occasion, and maybe our sort of sheep.

Not more than a few hundred sheep, all pure Llanwenog, were on show and scarcely more humans attended their sale. These were local sheep and local men, both bred and living in this one small area for generations. Hardly more than one exhibitor came from as far away as England, and the catalogue's list of past trophy winners indicated that the best Llanwenog sheep came from rival breeders who must be almost neighbours. It was all slow and easy here; our enthusiasm grew from pen to pen. We strolled about within the covered market, falling into conversation with smiling Welshmen who, it seemed to us, were so modest, so regional, so unambitious that they had no clear answer as to why their splendid Llanwenog sheep – as they seemed to us – were not more widely known and bred. One large quiet man knew for certain that they had been bred by his own family for over one hundred years, and another smilingly confessed that well, yes, perhaps they had been a little backward in selling the breed. These people were, as we had always found them, so cheerful, so helpful, so friendly that talking to them gave me the feeling that I really belonged among them, a feeling I had so desired but never achieved at a hundred events before.

One such family – father, son, daughter, son-in-law – stood chuckling and chatting by a pen of five ewe lambs so strikingly more attractive than all the others, so matching, so well-presented, that the rosettes and card which proclaimed them Champion Pen of Ewe Lambs was almost unnecessary. I immediately marked them in our catalogue as 'Beautiful' and when shortly we fell into natural conversation with the young son, I found myself assuring him through sheer high-spirited admiration that we fully intended buying them. Which we did. We also bought the largest and handsomest ram in the place, winner of the third prize in the open class, paying more money for him and the young ewes than could possibly be justified by any occasion other than this happy re-entry of ours into sheep keeping. Never mind; we were happy. Mr. Edwards and his family were happy too, slightly dazed by their show success, for they had never exhibited before, and they loaded their trailer with due loving care and followed us home to Dolgwili with the prize nucleus of our new flock.

"We simply cannot call him anything but Cassius," my glorying lady said when the Edwards had departed and we stood admiring the large and shapely ram, so black-faced and handsome with that top-knot of curly wool on his forehead, nibbling at Dolgwili grass with his harem of five immaculate wives nearby. To us, besotted again, they seemed too beautiful to be merely sheep. Perhaps they were unicorns, or some rare creature that had secretly survived in a closed Welsh valley. Just for now, we did not consider all the heartaches they would surely involve us in, the long hours of work, the worry. Just for now, it was enough that they were here, the sheep of Dolgwili, to run our lives once more.